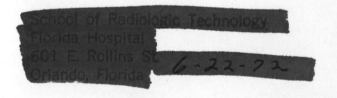

_Dr. W. C. Röntgen._

W. C. Röntgen.

*Dr. W. C. Röntgen.*

*By*

## OTTO GLASSER, PH.D., F.A.C.R. (Assoc.)

*Holder of Röntgen-Honor-Plaque of the Röntgen-Museum in Remscheid-Lennep, 1951. Professor of Biophysics, Frank E. Bunts Educational Institute Head, Department of Biophysics, Cleveland Clinic Foundation, Cleveland, Ohio.*

18082

*(Second Edition, Second Printing)*

Springfield · Illinois

## Charles C Thomas · Publisher

*Published and Distributed Throughout the World by*
CHARLES C THOMAS • PUBLISHER
Bannerstone House
301-327 East Lawrence Avenue, Springfield, Illinois, U.S.A.
Natchez Plantation House
735 North Atlantic Boulevard, Fort Lauderdale, Florida, U.S.A.

ISBN 0-398-02196-1

First Edition, 1945
Second Edition, First Printing, 1958
Second Edition, Second Printing, 1971
Second Edition, Second Printing, 1972

*With THOMAS BOOKS careful attention is given to all details of
manufacturing and design. It is the Publisher's desire to present
books that are satisfactory as to their physical qualities and
artistic possibilities and appropriate for their particular use.
THOMAS BOOKS will be true to those laws of quality that
assure a good name and good will.*

*Printed in the United States of America*
R-1

# Preface

The year 1945 provides an opportunity for celebrating a double anniversary in the scientific world. One hundred years ago on March 27, 1845 Charlotte Constanze Röntgen, wife of Friedrich Conrad Röntgen, a merchant living in Lennep, Germany, had a son. Fifty years later on November 8, 1895 this son, Wilhelm Conrad Röntgen, discovered *A New Kind of Rays* and made a fundamental contribution to science.

The discovery of roentgen rays stimulated physicists of the late nineties into feverish activity at a time when there seemed to be little more to be discovered in physical science. From the new and fascinating study of the roentgen rays, founded upon pure science, the science of roentgenology and subsequently of applied radiology developed. Other discoveries, following in rapid succession, included the electron, radioactivity, radium, and atomic disintegration. Innumerable new approaches to many scientific fields were revealed; countless lives in war and peace were saved.

The enthusiasm that hailed the use of roentgen rays from the publication of Röntgen's first communication has never abated. In gratification for this monumental gift to science, the American Roentgen Ray Society, the Radiological Society of North America, and the American College of Radiology act as sponsors of this commemorative volume, containing a new translation of the three classic papers on *A New Kind of Rays,* and dedicate it to the memory of the discoverer.

Charles C Thomas accepted the publication of the book with his characteristic enthusiasm. Virginia Thatcher and

Bernard Tautkins lent invaluable assistance in the writing and the illustrating of the manuscript. A factual basis for the material of the work is appended in a bibliography.

July, 1945                                    OTTO GLASSER

# Foreword to Second Edition

Increased interest in the life and work of the discoverer of x-rays is responsible for this second edition of *DR. W. C. RÖNTGEN*. It is, in essence, a reprint of the first edition since relatively few new points of view regarding Röntgen and the early history of his rays have meanwhile been brought forward. References to what new knowledge on the subject has been added are given on page 160. Some of the old illustrations have been replaced by better ones. The author acknowledges gratefully the continuous assistance of the Deutsches Röntgen-Museum in Röntgen's birthplace Remscheid-Lennep in furnishing new information and notably new illustrations on the history of roentgen rays.

OTTO GLASSER

# Contents

Dr. W.C. Röntgen.

## Chapter I

# 1845-1872

N THE SMALL PROVINCIAL TOWN of Lennep, lying in pleasant rolling country on the fringes of the Ruhr, yet, as evidenced by its many small industries, closely related to the industrial cities of the Ruhr valley, Wilhelm Conrad Röntgen was born on March 27, 1845 in a gray frame and slate house on Poststrasse. He was the only child of Friedrich Conrad Röntgen, a textile merchant, which was an unusual economy in the history of this family of solid and prolific burghers, who for many generations had lived in or near the town. Although most of his forbears had been coppersmiths or weavers, merchants or cloth manufacturers, exceptional deviations in the strain had brought forth a number of notable musicians, a gifted engineer, who ran the first steamboat to plow up the Rhine, and a master cabinetmaker, whose exquisite marquetry was to be found in the collections of Marie Antoinette, Louis XVI, Catherine of Russia, and Frederick William II of Prussia.

Wilhelm's mother, Charlotte Constanze Frowein, was his father's first cousin. Leaving the same town of Lennep several generations before, her ancestors had moved to Holland, where in shipping circles in the city of Amsterdam they were known as worthy and prosperous merchants and traders. Wilhelm's religious heritage was Lutheran and Reformed, and modifying the German and Dutch background were English, Italian, and Swiss strains from maternal grandparents from London, Turin, and Geneva.

That out of this rather ordinary and unpretentious background a genius should arise was nowhere indicated, particularly a genius in exploring the realm of pure science, yet undoubtedly of immeasurable value to Röntgen was the endowment of a century old industry and sobriety and pride in good honest workmanship. He also came by a fair share of Rhenish humor, frequently called forth to advantage in later years.

Until his travels as a young man took him to Germany, Wilhelm knew little of his native land, and the tall narrow slate house on Poststrasse, modeled for him as a toy by his father, was his only recollection of Lennep. For in May

"In this house was born, on March 27, 1845, the discoverer of the roentgen rays, Wilhelm Conrad Röntgen. His home town made him an honorary citizen in 1896 and erected this tablet on his seventy-fifth birthday, Lennep, March 27, 1920."

1848, when Wilhelm was three, the Röntgens moved to Apeldoorn in Holland, about one hundred miles to the northwest, where Charlotte's parents had made their home. Whether the Revolution of 1848 had anything to do with the emigration is not known, but it is a documented fact that the Röntgens lost their Prussian citizenship on May 23, 1848, and that they became Dutch citizens within a few months.

As the only child of a conservative and well-to-do merchant, Wilhelm had a pleasant childhood and was certainly indulged if not spoiled. Apeldoorn is a beautiful city, and to it noble families, who favored the surrounding countryside with their great estates, attracted an unusual culture and a rich art. A reflection of this culture was to be found in the Röntgen home on the Dorpstraat, in paintings by old Dutch masters, one a museum piece of the Holy Family, in beautiful silver, Wedgwood and Meissner china, and in art treasures brought from the Orient by an ancestor of the Froweins. In this environment Wilhelm developed a taste for fine art and well appointed living.

From the outset his schooling was erratic. After attending the primary and secondary public schools in Apeldoorn, he entered the private boarding school of Martinus Hermann van Doorn, situated in a rather large country estate on Middellaan. From that time nothing is known from available records about his schooling until he was registered in the Utrecht Technical School in December 1862. In Utrecht he lived with the Gunning family in the Nieuwegracht. Gunning was a well known chemist, and for him Wilhelm developed a healthy respect, with the result that submitting to a program of discipline in the Gunning home in no way interfered with his complete happiness.

The Technical School, which was suspended in 1866 with

the opening of the "hoogere burgerschool," prepared its students in two years for entrance into a technical high school, but did not fulfil the prerequisites for matriculation to a university. Here Wilhelm took courses in algebra, geometry, physics, chemistry, and technology. As a student he was only average. Making mechanical gadgets, riding or skating, or exploring the fields and forests of the surrounding countryside was, as he saw it, time well spent. He also had an appreciation for the ludicrous that eventually got him into trouble.

A fellow student, who was something of an artist, chose an unpopular teacher to caricature on the firescreen of the schoolroom. Not being gifted in drawing, Wilhelm greatly admired the ability of his friend and was savoring the full flavor of the debasing representation when his unrestrained enjoyment was cut short by the entrance of the schoolmaster himself, who in a rage demanded the name of the unflattering artist. When Wilhelm refused to reveal it, the director of the institution was summoned, and a series of painful cross examinations followed. In order that the enraged teacher should be pacified, it was inevitable that Wilhelm, too proud to reveal the name of the real culprit, should be expelled.

None of the Röntgens had been scholars, and only a few, an attorney and two wound surgeons, had received a university education. And although Friedrich secretly hoped that Wilhelm would succeed him as a successful merchant, it was important to him that his son should be equipped to enter a profession. When the unfortunate incident in the Utrecht school promised to be serious enough to interrupt Wilhelm's education, the father set out to obtain permission for a private examination for the *absolutorium*, which would give Wilhelm credentials to enter a college. After many months this permission was finally granted, and for almost a year Wilhelm prepared for the examination, notably by adding Latin and Greek to his studies. Then, the unforeseen happened. On the day before the examination the examiner, who was sympathetic toward Wilhelm, became ill and was replaced by a teacher who had taken part in the former suspension proceedings. With this handicap Wilhelm failed the examination. For the second time, and perhaps forever, the path to the university had been blocked. Yet it is doubtful that being deprived of formal classroom education was a real hindrance to him. It is rather probable that freedom from oppressing routine permitted the development of his capabilities in a way that proved invaluable to him in his subsequent career.

In order that he should at least be exposed to higher learning, he registered at the University of Utrecht in January 1865 with permission to audit courses as a student of philosophy, *privata institutione uses est,* and for two semesters attended lectures by Buys-Ballot on analysis and van Rees on physics, as well as others on chemistry, zoology, and botany.

By this time both Wilhelm and his parents had become resigned to his seeming inability to adjust to the requirements of the Dutch educational system and to obtain the credentials necessary to become a regular university student. Accordingly when a young Swiss named Thormann, who lived in Utrecht, informed them that the Polytechnical School in Zürich accepted after stiff entrance examinations students lacking the usual credentials, Wilhelm applied at once for admission. In his application he wrote, "Not satisfied with my course of studies at this university, I would

Röntgen (center) with his parents and other members of his
family (about 1865)

like to follow the call of the Zürich school and devote myself to applied mathematics. . . ."

A combination of favorable circumstances resulted in the examinations' being waived and in his being admitted as a student in the mechanical technical division of the Polytechnical School. When he arrived in Zürich on November 16, 1865, a few days after classes had started, he carried with him a letter from his physician explaining that he had been under treatment for an eye condition. This, in addition to his being two years older than the average student and being able to present good reports from the Utrecht schools, notably in mathematics, prompted the school authorities to be lenient and to accept him in good standing.

For three years he applied himself to the study of mathematics, technical drawing, mechanical technology, engineering, metallurgy, hydrology, thermodynamics, and many other branches of mechanical engineering. His instructors included some of the most prominent men in mechanical engineering and allied sciences, Christoffel, Prym, Zeuner, Kronauer, Ludewig, and Reye. His grades on the final examinations for his diploma as a mechanical engineer, received on August 6, 1868, were excellent.

Although he was becoming most adept in the exact methods of engineering sciences, he found himself more and more absorbed as time went on by basic sciences, permitting delving into realms of the unknown. He saw clearly that a knowledge of them, combined with logical reasoning and accurate experimentation and deduction, offered great opportunities for wresting unknown secrets from nature. At that time choosing one of them to master did not occur to him, and he had no foreknowledge that physics would be his ultimate choice. Until receiving his diploma in engineer-

ing, Röntgen had but one course in physics, that given by
the famous Clausius, the father of thermodynamics, on
technical physics during his second semester at Zürich. It is
an arresting fact that Röntgen, the great experimental
physicist, never had a fundamental course in experimental
physics in college.

Much of the credit for Röntgen's eventually applying
himself to the study of physics must be given to August
Kundt, who in 1868 succeeded Clausius in the chair of
physics at the Polytechnical School. After graduating from
the school in engineering, Röntgen stayed on in Zürich to
take a few more courses in mathematics and also Kundt's
lecture on the theory of light. In Kundt's laboratory he
undertook his early physical experiments on various properties of gases, and he gradually came to the realization that
this was the work for which he was suited. After a year
of association with Kundt he submitted his thesis, *Studies
on Gases,* to the University of Zürich, housed in the same
building as the Polytechnical School, and on June 22, 1869
obtained his doctorate in philosophy from that university.

This creditable conclusion of his education in Zürich did
not entirely reflect his college career; not overnight did he
change from a fun-loving student into a serious scholar.
Zürich is a wonderful place in which to be a student if one
has sufficient money and not too strong an ambition. Above
the city to the south rises the Uetliberg, and sharing with
the city the lap of the great mountain chain is beautiful
Lake Zürich. Of the many paths to the summit of the
Uetliberg, Wilhelm had a particular fondness for the
*Leiterli,* although once on this steep trail he slipped in
the soft clay and falling into a quarry was severely injured.
During those years in Zürich there developed in him a
love for mountain-climbing and for that unparalleled

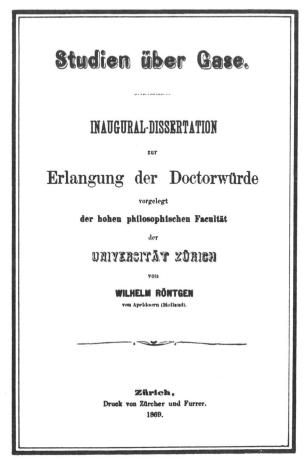

Studien über Gase.

INAUGURAL-DISSERTATION

zur

Erlangung der Doctorwürde

vorgelegt

der hohen philosophischen Facultät

der

UNIVERSITÄT ZÜRICH

von

WILHELM RÖNTGEN

von Apeldoorn (Holland).

Zürich,
Druck von Zürcher und Furrer.
1869.

Title page of Röntgen's thesis, *Studies on Gases*, submitted to the
University of Zürich for the degree of Ph.D.

serenity of the majestic snowcapped Swiss mountains that
drew him back to them year after year throughout his whole
life. And then there were boating parties and gatherings
with congenial friends in the inns. In any crowd of students
Wilhelm was outstanding. He was tall and handsome and

looked out at the world with clear penetrating eyes, and in the fashionable attire of the day—dark coat, gray trousers, soft collar with big wing tie, and gold watchchain around the neck—he was fair competition for the affections of any young Zürich woman.

As long as he was in Zürich, Röntgen lived in a rented room on Seilergraben, No. 7, and ate regularly with several students and unmarried professors at an old restaurant, *Zunft zur Waag*. However, the favorite inn of the students, and one not far from the Polytechnical School, was *Zum Grünen Glas*, where singing and drinking were mixed with ardent conversation, often stimulated by the innkeeper, Johann Gottfried Ludwig. Johann had been forced to flee Jena for his revolutionary activities in the thirties and consequently had been barred not only from his university but also from his country. He was a man of no mean intellect and scholarly attainment. Many prominent German refugees had fled to Switzerland at that time, and some, the chemist Bolley, the philosopher and author Vischer, and the poet and historian of art Kinkel, had found posts in the Polytechnical School. Johann, however, had been forced to look elsewhere for employment and had taken to innkeeping as a way of making a living. Eventually he had married a young Swiss woman, Elisabeth Gschwend, by whom he had three daughters. As an avocation that allowed him continued contact with university life, he taught the students fencing, tutored them for examinations, and sometimes translated a thesis into Latin, which was still a requirement at the university.

Between the "Poly" student Röntgen and the innkeeper a strong liking developed, based upon a mutually free exchange of ideas. Johann's affection for Wilhelm was shared by his second daughter, Anna Bertha, and general conversa-

tions in the inn led to private words between the young
woman and the student. Bertha, who was six years older
than Wilhelm and a tall slender girl of extraordinary charm,
may have flattered him by her interest as an older woman
and by her obvious favoritism. Gradually the tenor of
their conversations changed from pleasantries and anecdotes
about mountain-climbing and rowing on the lake to serious
discussions about his plans for the future.

A letter to Bertha from her father while she was in a
girls' boarding school in Neuchâtel in the fifties shows her
stripe and pretty much what was expected of her:

I do not want to lay aside the pen with which I have just
written to your teacher without also writing you a few words.
I am very happy to learn from Miss Grossmann that she is
well satisfied with you in every respect. This shows that you
appreciate the great sacrifice which your parents are happy to
make for your education and that you are determined to con-
tribute all you can to become what we would like you to be,
sincere, orderly, and morally and scientifically well educated.
These are the only treasures that you can acquire with the
help of our efforts and those of your good teachers. Other
treasures we cannot give to you, and even if we could, they
truly would not be worth as much as these.

I was happy to read that you continue to feel well. You
seem to like the climate and mode of life, which makes you
long less for your parents' home. You may easily forget the
pleasures you shared with us. The weather continues to be bad
and takes away all of the summer pleasures; it probably is the
same in Neuchâtel. And now, my dear daughter, continue to
be industrious and good, so that we may always receive good
news of you. A very sincere good-by from all of us—Your
loving father.

When Wilhelm received his doctorate in philosophy
from the University of Zürich, Bertha was spending several

months in a sanatorium on the Uetliberg for her health and at the same time was profitably indulging her hobby of collecting plants and flowers. On the evening of the graduation day his one thought was to share his joy in achievement with her, and he quickly climbed the *Leiterli*. Whatever that evening's expression of their happiness, it is certain that they made optimistic plans about the future, although to Röntgen two degrees and a possible assistantship in Kundt's physical laboratory were in no sense a guarantee that he would be able to support a family. Wilhelm's father reported further developments in a letter to Mr. Buscher on October 3, 1869:

. . . After we left Lennep we continued our trip for three short days and arrived by railroad in Zürich, where we were greeted by our son. We were very happy to be with him and met a Zürich girl about whom Wilhelm had already written to us. We had not given a definite answer to his letters, but when he insisted on having our opinion, we considered it our parental duty to look into the matter and were favorably impressed when we met her. Then we spent two weeks in Zürich and decided that, in order to get better acquainted with the girl, we would take her and Wilhelm to Baden-Baden for a few days and then to Wildbad for two weeks. The result was that when we said good-by in Karlsruhe, for we were returning home and they to Zürich, we gave our consent to the engagement. The girl (Bertha Ludwig) is well educated, comes from a good family, is intelligent, of good character, and very agreeable. . . .

Meanwhile Kundt had recognized in young Röntgen the qualities of a good physicist and had asked him to become his assistant. In helping Kundt organize an experimental physics laboratory, despite being restricted by inadequate quarters and poor equipment, he learned a great deal and found his assistance helpful to two advanced students, Franz Exner and H. Schneebeli. The three remained friends for life. Exner later became professor of physics at Vienna, and Schneebeli held the same post at the Zürich Polytechnical School.

The relationship between Röntgen and Kundt was excellent, and they had but one serious disagreement during their many years of collaboration. In one room of the institute Kundt kept some especially fine instruments and glassware, which he absolutely forbade any other person to touch. Röntgen dared to ignore this prohibition and as a result was caught in this sanctuary one Sunday afternoon by Kundt himself. The fiery tempers of the two men clashed. Soon, however, Kundt became convinced of Röntgen's honesty, and the episode was forgotten.

In 1870 when Kundt was called to the chair of physics at the University of Würzburg, that ancient city in the vine-clad valley of the Main, he took his assistant Röntgen with him. Again Röntgen found a poorly equipped physical institute in the old university building on Neubaustrasse, but he attacked his various duties and undertook research problems with youthful vigor. During this time he lived in an inn, *Eckert's Garden*, in Veitshöchheimerstrasse.

Meanwhile Bertha spent many months in Apeldoorn to learn from mother Röntgen the fine art of German and Dutch cooking. This separation of the young persons increased their eagerness to be married. The sumptuous ceremony took place at the Röntgen home in Apeldoorn on January 19, 1872. The young Swiss Thormann was the best man, and favorite guests were Wilhelm's uncle Ferdinand and cousin Louise. Returning to Würzburg, Wilhelm and his bride moved into a modest house on Heidingsfelderstrasse.

### Huwelijks=Acte
van
### Wilhelm Conrad Röntgen en Anna Bertha Ludwig.

Op heden den negentienden Januari des jaars achtien
honderd twee=en=zeventig, compareerden voor Ons Meester
**Pieter Marius Tutein Nolthenius**, Burgemeester, Ambtenaar
van den Burgerlijken Stand der Gemeente **Apeldoorn**, Provincie
Gelderland,

............ **Wilhelm Conrad Röntgen,** jongman,............
oud zes en twintig jaren, van beroep assistent aan het physicalisch
laboratorium geboren te **Lennep** en wonende te **Würzburg,**
meerderjarige Zoon van de echtelieden **Friedrich Conrad Röntgen**
oud een en zeventig jaren, zonder beroep, en **Charlotte Constance
Frowein,** oud vijf en zestig jaren, zonder beroep, beide te **Apel=
doorn** wonende, alhier tegenwoordig en in dit huwelijk toestem=
mende, ...................................................;.
...................... En ......................
............ **Anna Bertha Ludwig,** jonge dochter,............
oud twee en dertig jaren, zonder beroep, geboren te **Zürich,** en
wonende thans te **Apeldoorn,** vroeger te **Schwamendingen,**
meerderjarige dochter van de echtelieden **Johan Gottfried
Ludwig,** overleden, en **Elisabeth Gschwend,** zonder beroep, te
**Zürich** wonende welke Comparanten Ons, in tegenwoordigheid
van de vier hierna genoemde getuigen hebben verzocht, om over
te gaan tot de voltrekking van het door hen voorgenomen
Huwelijk.

En hebben wij, voornoemde Ambtenaar, aan dit verzoek
voldoende:

Gezien de nagemelde aan Ons overgelegde stukken.

10. Extract uit de geboorte=acte van den bruidegom;
20. Bewijs van voldoening aan de Nationale Militie;
30. Extract uit de begoorte=act der bruid;
40. Twee bewijzen van te **Würzburg** en te **Schwamendingen**
ergane huwelijks=affondigingen.

Gelet de Acte der Huwelijks=affondigingen gedaan zijn te
**Apeldoorn** op de Zondagen van den zevenden en van den
veertienden Januari dezes jaars en te **Würzburg** en te
**Schwamendingen** op den zevenden Januari dezes jaars.

In aanmerking nemende dat geene stuiting van dit Huwelijk
ter Onzer kennisse gekomen is:

Gehoord de verklaring der partijen, dat zij elkander tot
echtgenooten aannemen, en dat zij getrouwelijk, al de pligten
zullen vervullen, welke door de wet aan den Huwelijken Staat
verbonden zijn:

Verklaard in naam der wet dat **Wilhelm Conrad
Röntgen,** en     **Anna Bertha Ludwig,**
door het Huwelijk zijn verbonden.

Waarvan wij beze Acte hebben opgemaakt in tegenwoordig=
heid van **Richard Röntgen**, oud zestig jaren, zonder beroep,
wonende te Velp, oom van den bruidegom, **Jacob Bobbens**, oud
brie en bertig jaren, zonder beroep, wonende te Apeldoorn,
**Wilhelm Walter**, oud zes en bertig jaren, van beroep notaris,
wonende te Apeldoorn en **Carl Ludwig Wilhelm Thormann**,
zonder beroep, oud zes en twintig jaren, wonende te **Jsselstein**,
de beide laatsten aan partijen in bloed= of aanverwantschap, niet
bestaande, uitbrukkelijk verzochte getuigen, welke, na voreliging,
met ons den bruidgom, de bruid en de ouders van den bruide=
gom hebben geteekend.

                                          (get.) **Lutein Rolthenius**
(get.) Dr. W. C. Röntgen
        A. V. Ludwig
        Fr. Conr. Röntgen
        C. C. Röntgen=Frowein
        Richard Röntgen
        J. Bobbens
        W. Walter
        C. L. W. Thormann

The glow of their happiness together is reflected in
letters written later in the same year to Louise, who with
her father was preparing to emigrate to the United States.
From Strassburg on May 20 Bertha wrote:

I also wish to write you a few lines before you leave your
beloved country and everything in it that is dear and precious
to you.

I really should begin this little letter with an apology, but
since I know my dear Louise so well, I feel certain that she
will forgive my long silence, especially if I explain to her that
young, happily married people have too little time to write. . . .
Yes, dear Louise, it is infinite happiness to share with the man
one loves from the bottom of one's heart everything, happiness
and pain, everything that the dear, all-wise God sends us. And
then, dear Louise, to have one's own household, in which one
can do whatever one likes! Unfortunately, I am unable to
describe to you this happiness in words, but I shall pray to God
that you also may enjoy the same thing. . . .

And from Wilhelm on the same day:

Little did I suspect a few months ago when we celebrated my wedding so hale and happily that it would be the last time that we would see each other for an indefinitely long period. It is good that we cannot look into the future, for had I foreseen this event at that time, my great happiness and joy on my wedding day would not have been unmarred. These few lines will tell you, dear uncle and Louise, how often our thoughts have been with you during these last few days, and how we go with you through the painful and sad separation from uncle Richard and from my father. But these greetings shall also bring you a sincere good luck and God bless you on your long voyage; our very best wishes are with you. We hope and we know that your expectations for your future, which let you make this serious decision, will be fulfilled. May German love, industry, faith, and custom have their beneficial influence upon you and your surroundings also in the far West. Then you will soon send us reassuring and happy news, tell us that you are well in your new field of activity. Think once in awhile of those who hate to see you go, and always remember us kindly. A pleasant voyage, and all good wishes in your new home.

On the other hand, having been spoiled by his parents and having enjoyed for a number of years the freedom of independent living, Wilhelm was not easily domesticated. In their early years of marriage they had little money, and Bertha was obliged to do her own cooking, washing, sewing, and mending. Adjustments were sometimes difficult. Once a domestic argument reached such a pitch on the street that Wilhelm hailed a passing cab and sent his wife home alone. But they were young, and they overcame their temporary differences quickly and with good humor.

# Chapter II

# 1872-1895

 HE DECISION to follow a university career as a physicist was not, as Röntgen soon found, without impediments. Even being able to work with Kundt did not quite compensate for certain drawbacks in his affiliation with Würzburg University. First of all, he was beset by the countless irritations consequent to working in a meagerly equipped "physical cabinet." This, however, was secondary to the discovery that he was again being held back by his inadequate formal education. Before he could be appointed to a salaried position on the faculty, it was necessary that he climb the first step of the academic ladder and be appointed a *privat-dozent,* or unpaid lecturer recognized by the university. Not being able to present the *absolutorium* and lacking requisite training in the classical languages, he was prevented by the strict traditions of the old institution from getting this initial academic title.

Fortunately this temporary disquietude was relieved in April 1872, a few months after Röntgen's marriage, when Kundt took his assistant with him to the recently reorganized Kaiser-Wilhelms-University of Strassburg in the medieval fortressed capital of Alsace-Lorain. This university, founded in 1567 and having been practically inactive since the French Revolution, had been reopened after the Franco-Prussian War and occupied handsome, well appointed buildings. Here the Röntgens found a delightfully youthful and liberal atmosphere free from hampering traditions. After two years of conscientious and hard work Röntgen was appointed a *privat-dozent* on March 13, 1874. From that time his devotion to physics was equalled by an interest in teaching, and in addition to conducting experiments in fundamental researches he constructed a number of simple devices to demonstrate basic physical phenomena in his lecture courses.

With the brakes on his university career released, his ambition knew no bounds. Consequently when, on the basis of some investigations he had made with Kundt, an offer of a full professorship in physics and mathematics came from the Agricultural Academy in Hohenheim, Württemberg, in April 1875, he accepted, even though this meant breaking strong ties—those with Kundt, his parents, who had moved to Strassburg in 1873, and good friends with whom he and Bertha had explored the Rhine country, the neighboring Vosges mountains, and the Black Forest.

However, in Hohenheim he found facilities even more inadequate than those in Würzburg, and he soon learned that a full professorship did not compensate for his dissatisfaction in being unable to do creditable research. He and Bertha missed their Strassburg friends and not infrequently had to cope with rats trying to take over their apartment. Without hesitation a year and a half later he accepted an offer to return to the University of Strassburg as an associate professor in theoretical physics.

Those Strassburg and Hohenheim years were ones of perfecting technics in physical methods and gaining teaching experience. Going more deeply into the work on gases begun in Kundt's Zürich laboratory, he learned in studying the ratio of their specific heats an appreciation for accuracy in his own experimental results. He also wrote in collaboration with Kundt four important papers on their successful measurement of the effect of the electromagnetic rotation of the plane of polarization in gases. They not only proved the existence of this rotation but measured it quantitatively. This phenomenon their great predecessor Faraday had vainly attemped to demonstrate.

Working with crystals was always a joy to Röntgen, because he believed they were the embodiment of natural laws. With an unusual experimental resourcefulness he attemped to wrest from nature physical facts leading to such laws. Little did he know then that almost forty years later the riddle of the nature of his x-rays would be answered by studies with crystals. In Strassburg he investigated the conduction of heat in crystals and later extended his studies to include their actinoelectric and piezoelectric properties. His conclusions from the latter studies led to lively discussions with Kundt.

Among the fifteen papers published in his Strassburg period, between his twenty-fifth and thirty-fourth years, were two on discharges of electricity through conductors and insulators and one in collaboration with Exner, his old Zürich friend, on the determination of sun radiation with an ice calorimeter. He also published two reports on the theory of elasticity and capillarity and a number of shorter notes, *On Soldering Platinum-Plated Glasses, On an Aneroid Barometer with Mirror and Scale,* and *A Telephonic Alarm,* which demonstrated his technical skill.

In all these publications Röntgen proved himself an ideal pupil of Kundt as a classical experimental physicist. He clearly conceived the problem, was skilful in its experimental investigation, and carefully carried out rigid control tests of the results obtained before presenting his findings in a brief but precise and logical manner.

It was therefore not surprising that in 1879 he was recommended for the chair of physics at the Hessian university at Giessen by such great leaders in his field as von Helmholtz, Kirchhoff, and Meyer; this appointment he accepted. In Giessen the Röntgens made friends that were to last a lifetime, von Hippel the ophthalmologist, Hofmeier the gynecologist, Zehnder the physicist, and others, with whom in the spring or at the time of the vine festivals in the fall they made up parties for excursions to the Rhine.

In Giessen Röntgen continued and concluded investigations of the so-called Kerr effect begun in Strassburg even before this effect was discovered. Then he devoted several years to further studies of the properties of crystals. The effects of heat still fascinated him, and upon finishing his studies with crystals, he started a series of investigations on the absorption of heat rays in water vapor, an old problem that had led to intense arguments between John Tyndall and Gustav Magnus. With a simple homemade but very sensitive air thermometer, he proved that humid air is heated more quickly than dry air, in other words, that water vapor actually absorbs heat radiation.

Upon his arrival in Giessen he had found that the laboratories and lecture room, housed in the private home of his predecessor Buff on Frankfurterstrasse, did not compare favorably with those in Strassburg. Consequently he set out, with the support of his assistant Schneider and later of Zehnder, to improve the facilities for research. In 1880 a new physics laboratory and lecture room were built in connection with the university buildings, and during the winter term 1880-81 he took occupancy of this new institute.

His skill as a teacher and a lecturer increased. He wrote additional papers on demonstration apparatus that he designed and built himself, one of these, describing a lecture demonstration of Poiseuille's law, bringing especially favorable comment. In time he developed a skill for measuring very small physical effects with extreme accuracy, with the result that many of his findings remain unchallenged to this day.

Wishing to spend their last years with their children, Röntgen's parents had followed them from Strassburg to Giessen. Both died during those years, his father in 1884 in Giessen, and his mother in 1888 in Bad Nauheim, having been thoughtfully and affectionately cared for until the end. His mother's death moved Röntgen deeply. Her complete understanding of many of his problems had made him feel "one heart and one soul" with her. Years later he wrote to Zehnder, "The question how my mother would have reacted or spoken to me in this or that difficult situation has often led me to the right solution."

In 1887, being still childless after many years of marriage, he and Bertha decided to take Bertha's six year old niece, Josephine Bertha, into their home. The child was frail, and to counteract Mrs. Röntgen's indulgence of her whims Röntgen was frequently severe with her and subjected her to a strict discipline. She was legally adopted at the age of twenty-one. Also in Giessen Röntgen became financially independent for the first time. Almost the first display of affluence was the renting of hunting grounds in the Rimparer Forest, on which a small cottage was built to lodge hunting parties.

In 1888 Röntgen made his reputation on investigations proving that magnetic effects are produced in a dielectric, such as a glass plate, when it is moved between two electrically charged condenser plates. In this work, based upon the theoretical reasoning of the Faraday-Maxwell electromagnetic hypothesis, were ideally coordinated his ability

as a theorist and his genius in experimentation. Years later
he explained to Margret Boveri the distinction between
experimental and theoretical physics:

To my view there are two methods of research, the apparatus
and the calculation. Whoever prefers the first method is an
experimenter; otherwise, he is a mathematical physicist. Both
of them set up theories and hypotheses. . . .

Because of this synthesis between the theoretical conception of the problem and its experimental proof, the discovery of the "roentgen current," as it was named by his
colleague H. A. Lorentz, was in later years regarded by
Röntgen as being as important as his discovery of x-rays.
Sommerfeld, the theoretical physicist, later stated: "The
roentgen current, together with the Rowland effect, forms
an indispensable foundation for the conception that the
dielectric properties of matter depend upon the presence of
charges (electrons), and in its later quantitive perfection, it
decides directly against the original Maxwell-Hertz
theory." Satisfying recognition of the fundamentally important work came from many colleagues, notably from
von Helmholtz, who submited the paper to the Royal
Prussian Academy of Sciences in Berlin in 1888 and recommended ·it for publication.

To his colleague Hertz, who had sent him reprints of
publications on his epoch-making work on electric oscilla-
tions, Röntgen wrote on March 1, 1888:

Along with my thanks for your reprints, I send you my sincere congratulations on the excellent investigations that you
have made. I think that they are among the best made in physical science during the last years. I sometimes become impatient
with my own experiments and must then begin all over again.
I must postpone detailed publications of them, but have pre-

sented a brief report, which I enclose, on the electrical effects of moving dielectrics.

Little did Röntgen then suspect that Hertz's work would lead him to a great discovery, which Hertz himself would not live to see.

As a consequence of his investigations Röntgen received other kinds of recognition. In 1886 the University of Jena called him to its chair of physics, and two years later the University of Utrecht asked him to succeed Buys-Ballot, his old teacher, in a similar position because of "the quality of your publications, which show an exceptional intellect and profound knowledge, combined with originality of ideas." In the same vein he was called a "brilliant teacher of great experimental skill." Röntgen declined both offers, and, not without a certain satisfaction at the turn of fate that brought him an offer from the university that had rejected him as a regular student, he wrote in reply to the Utrecht call, "To move into entirely new surroundings would require too much of my time, which I had rather devote to scientific investigations. . . ."

That Röntgen was able to work successfully on so many problems in different branches of physics was due to his tremendous knowledge of the literature, and often until late into the night he read in order to keep himself informed on current publications. His devoted application to the study of physics and his good honest workmanship in the research laboratory produced excellent results. Only with the greatest effort, however, could he force himself to write down his observations for publication. His interest was quickly transferred to other experiments, and often he became so absorbed in these that he did not care about publishing the results of earlier investigations.

During the Giessen period, between his thirty-fourth and forty-third years, he published eighteen papers. In addition to the studies already mentioned, there were several collaborative efforts with Schneider on the compressibility and surface tension of a number of liquids and with Zehnder on the influence of pressure upon refraction coefficients of several liquids.

On October 1, 1888, when Röntgen was forty-three, an offer came that he could hardly decline. Friedrich Kohlrausch, the great experimental physicist of Würzburg, went to Strassburg, and Röntgen was offered Kohlrausch's post as professor of physics and director of the new physical institute of the University of Würzburg. His accurate experimental methods of investigation made him an ideal successor to the master of the technic of physical mensuration.

It was a far cry from the old, meagerly equipped "physical cabinet" of the early eighties to the new physical institute on the broad, tree-lined Pleicher Ring, later to be known as Röntgen Ring. The building had two spacious floors, a basement, and a lecture room. The second floor comprised the private residential quarters of the director of the institute and had ample room for a conservatory, much to Mrs. Röntgen's delight.

The Röntgens' return to Würzburg was something of a triumph. To this city sixteen years before they had come as newlyweds, uncertain but optimistic about the future and deeply disappointed when the university refused to give Röntgen an academic title. Now at the height of their maturity, with Röntgen's reputation as a physicist firmly established, they were quickly accepted into the inner circle of the university and enjoyed a vivid social life with a large group of friends—Boveri the biologist, Kölliker

"In this building W. C. Röntgen discovered in the year 1895 the rays which are called by his name."

the father of histology, Fick the physiologist, Stöhr the anatomist, Sachs the botanist, Kunkel the pharmacologist, Prym the mathematician, Tafel the chemist, Leube the specialist in diseases of the stomach, and many others.

To Louise, his cousin in the United States, he wrote on December 30, 1890:

As you have sufficiently experienced, I am really a bad correspondent, but you must believe me when I say that I frequently think of you with love. Since we do not write, we do not hear anything from you, and still I should like to know how you all are and especially you. I console myself continually with "no news, good news." However, now at the end of the year I have a very urgent desire to learn how you are, and therefore I finally sit down to write you this letter.

Starting with news from our side, I can generally report only good things. Bertha's health has not been the best for several years, only rarely does she really feel completely healthy and fresh. However, we consider it a great fortune that her spirits and her humor do not suffer. I myself feel well always and enjoy life with the exception of times when I am somewhat nervous from too intensive work. The position that I occupy now really makes me very grateful and happy; frequently I wish my dear parents were still here to enjoy the results of their efforts and troubles! Who of us would have thought twenty-five years ago that I once would occupy a professorship in one of the larger universities of Germany? Indeed, I have had good luck in my life.

Our little household has been somewhat enlarged. A young friend from Munich lives with us and will probably stay a year with us. We also have taken into our house a little niece of my wife's, a child nine years old. It will depend upon the success of our education whether we shall keep her or not; as already stated, the change in our household made last fall is a temporary one.

We have celebrated Christmas in the usual way. The big

tree was beautifully lighted. The Christ Kind had brought a number of presents. Do you still remember the windmill that was run by sand, the fountain, and the crèche that my father had made in Apeldoorn for Christmas? At that time according to the Rhenish custom Christmas was celebrated on the first Christmas day, very early in the morning; then one went to church.

Would you please use the enclosed sum to make your children happy, and tell them about their far distant cousin in Germany.

If you write to Heinrich, send him my best regards. With sincere regards from the two of us to you.

In his first six years at the university Röntgen published seventeen papers. He obtained important results in studies on influences of pressures upon various physical properties of solids and liquids, some of the work having been done in collaboration with Zehnder. He investigated the compressibility of many liquids, notably ether and alcohols, and continued studies of pressure upon the dielectric constant of water and ethyl alcohol. He examined the refractive index of these liquids and the conductivity of various electrolytes. New studies initiated included the measurement of the thickness of coherent layers of oil on water, from which conclusions were drawn about the "radius of effective spheres" of the molecule and the range of diameter of molecules. He then coordinated his results and those of other investigators on the compressibility of water. From the abnormal action of water as compared with other liquids, that is, its decrease of compressibility with increasing temperature and decrease of internal friction through pressure, or through the increase of the thermal coefficient of expansion with pressure, he concluded that water consists of two types of molecules. These two types are, first, ice molecules, which cause a larger volume, and, secondly, molecules formed with an increase in temperature with a consequent decrease in volume. In addition to these important researches, Röntgen again proved himself a master of experimentation and demonstrated his mechanical skill in several descriptions including *Some Lecture Demonstrations, Methods of Producing Pure Surfaces of Water and Mercury*, and *A Note on a Method of Measuring Differences in Pressure by Means of Mirror and Scale.*

That Röntgen's interests were not entirely restricted to his activities in the physical institute is evidenced by the fact that in 1894 he was elected to the rectorship of the university. In assuming this highest office bestowed by the university he said:

The university is a nursery of scientific research and mental education, a place for the cultivation of ideals for students as well as for teachers. Its significance as such is much greater than its practical value, and for this reason one should make an effort in filling vacant places to choose men who have distinguished themselves not only as teachers but as investigators and promoters of science; for every genuine scientist, whatever his field, who takes his task seriously, fundamentally follows purely ideal goals and is an idealist in the best sense of the word. Teachers and students of the university should consider it a great honor to be members of this organization. Pride in one's profession is demanded, but not professorial conceit, exclusiveness, or academic presumptuousness, all of which grow from a false estimation of one's self, rather a vital feeling of belonging to a favored profession, which gives many rights but also requires many duties. All our ambitions should be directed toward a faithful fulfilment of duties toward others as well as toward ourselves—only then will our university be esteemed, only then shall we prove worthy of the profession of academic freedom, and only then will this valuable and indispensable gift be retained. . . .

And then proceeding with that which was closest to his heart:

. . . Only gradually has the conviction gained importance that the experiment is the most powerful and most reliable lever enabling us to extract secrets from nature, and that the experiment must constitute the final judgment as to whether a hypothesis should be retained or be discarded. It is almost always possible to compare the results of ratiocination with practical reality, and this gives the experimental research worker the required assurance. If the result does not agree with reality, it must necessarily be wrong, even though the speculations that led to it may have been highly ingenious. Perhaps one may see in this necessity a certain inexorability, when one considers the great mental effort and the great amount of time required in the accomplishment of the result and the many fond hopes that must be destroyed in the process. Yet the investigator in natural sciences is fortunate to have such a touchstone, even though it sometimes brings great disappointments.

During the year 1894 three colleagues died for whom Röntgen had the greatest admiration. On January 1 Heinrich Hertz, one of the most talented and original physicists of his time, then a professor at the University of Bonn, died at the age of thirty-seven. On May 21 August Kundt, Röntgen's beloved preceptor and at that time professor of physics at the University of Berlin, died at his country home near Lübeck. And on September 8 the nestor of German science and notably physics, Hermann von Helmholtz, President of the Physikalisch-Technische Reichsanstalt in Berlin-Charlottenburg, passed away. The loss to science seemed irreparable. Röntgen's personal loss was more deep because of the indebtedness he felt toward each of them for invaluable help in his own career.

Early in 1895 he received a call from the University of Freiburg i. Br., to become professor of physics and director of a new institute. The offer was attractive, and the Röntgens were tempted to move to the beautiful city at the foot of the Black Forest mountains. But after most careful consideration, Röntgen decided to stay in Würzburg.

Now in his fiftieth year Röntgen could justifiably have considered his previous work creditable and his situation in life satisfying. Yet some time in October of that year 1895, with the ever youthful curiosity of a scientist, he found his entire attention captivated by the work of Hittorf and Crookes and of Hertz and Lenard on cathode rays. In June of the previous year he had procured an excellent Lenard tube from the glassblower Müller-Unkel in Braunschweig and had repeated some of Lenard's original experiments, notably his observations on effects produced by cathode rays in free air and hydrogen. These beautiful experiments occupied his interest for some time and finally became so absorbing that he decided to drop further studies on the influence of pressure upon dielectric constants of various liquids and to devote himself exclusively to researches on cathode rays.

## Chapter III

# November 1895

ITH WINDOWS overlooking the garden of the physics institute toward the east, two laboratories at the end of a long hallway on the first floor were selected by Röntgen for the cathode ray experiments. Into the larger of the laboratories was moved a Ruhmkorff induction coil of the type manufactured by Reiniger-Gebbert and Schall in Erlangen. This was equipped with a Deprez interrupter and produced sparks from four to six inches in length. Several Hittorf-Crookes tubes, or discharge apparatus as Röntgen preferred to call them, were arranged on a shelf, together with some Lenard tubes, including the one he had obtained the year before from Müller-Unkel. Other equipment, notably a Raps vacuum pump, was then reassembled to make and evacuate additional Hittorf-Crookes and Lenard tubes.

The particular apparatus that Röntgen proposed to use in the cathode ray experiments was based not only upon the production of high tension electric charges but upon their conduction through highly evacuated vessels. The knowledge of electricity had had its origin almost three centuries before in *Gilbert's* studies of magnetism and static electricity. Subsequently simple demonstrations of electrical phenomena had had wide public appeal, particularly those of the effects of static electricity upon evacuated spaces and living organisms. Eventually, largely through the efforts of scientists such as *Franklin, Galvani, Volta, Am-*

*père, Ohm, Faraday,* and *Henry,* the knowledge of electricity had been systematized.

The barometric vacuum had been discovered by the Italian *Torricelli* in 1643, and about the same time *Otto von Guericke,* a German burgomaster, had built an air pump with which he could slowly evacuate air from enclosed vessels. Vacuum pumps and simple electrostatic machines had been further developed by the English scientists *Robert Hooke* and *Robert Boyle* in 1660 and by *Francis Hauksbee* in 1705. The precursor of Röntgen's own tubes for the cathode ray experiments was doubtless that developed by *Heinrich Geissler.* About 1855 Geissler, a glassblower at the University of Bonn, had built a practical mercury vacuum pump with which he had evacuated his famous tubes. These tubes filled with various gases showed beautiful colored effects when high tension discharges from an induction coil were passed through them. *Plücker,* professor of physics at the University of Bonn and Geissler's chief, had helped in the refinement of the tubes, and his historic observation of an emanation from the cathode end of the tubes, when a high tension discharge was passed through them, had been among the earliest to be made of the so-called cathode rays.

To *William Crookes* these "cathodic" rays had represented a new "fourth state of matter," and one by one their real properties had been explored by workers such as *Hittorf, Hertz, Goldstein,* and *Lenard.* In order to study the properties of these mysterious cathode rays outside the tube in free air, Hittorf and later Lenard had equipped ordinary glass cathode ray tubes with extremely thin aluminum windows, through which the cathode rays could penetrate to the outside. In a series of classical experiments Lenard had found that the rays made the air electrically conductive, but that they were easily absorbed in a few centimeters of free air. He had also discovered that they produced luminescent effects upon certain fluorescent salts and darkened the photographic plate.

Many of these experiments had been repeated by Röntgen, and now in initiating further researches he suspected that the theoretical speculations of *Maxwell* and *von Helmholtz* on electric and magnetic disturbances within the so-called ether might also be of significance in the interpretation of some of the observed phenomena.

In repeating Lenard's experiments, Röntgen, according to the inventor's suggestion, enclosed the Lenard tube in a tightly fitting cardboard coat covered with tinfoil, which evidently had been supplied for the purpose of protecting the thin aluminum window of the tube from possible damage in the strong electrostatic field, but which at the same time prevented any visible light from the tube from penetrating to the outside. He again was able to confirm by his own observations that invisible cathode rays emanated from the tube and did produce a fluorescent effect on a small cardboard screen painted with barium platinocyanide, but only when this screen was placed fairly close to the window. Now it occurred to him that in similar experiments with heavier-walled Hittorf-Crookes tubes fluorescence of such a screen might also be caused by cathode rays, but that it might possibly be obscured by the strong luminescence of the excited tube.

This idea fascinated Röntgen. Late one afternoon when, as was his custom and preference, he was working alone in the laboratory, he determined to test the ability of a Hittorf-Crookes tube, that is, an all glass tube without a thin window, to produce fluorescence on the barium platinocyanide screen. Selecting a pear-shaped tube from the rack, he covered it with pieces of black cardboard, carefully cut and pasted together to make a jacket similar to the one used previously on the Lenard tube, and then hooked the tube onto the electrodes of the Ruhmkorff coil. After darkening the room in order to test the opacity of the black paper cover, he started the induction coil and passed a high tension discharge through the tube. To his satisfaction no light penetrated the cardboard cover.

He was prepared to interrupt the current to set up the screen for the crucial experiment when suddenly, about a yard from the tube, he saw a weak light that shimmered on a little bench he knew was located nearby. It was as though a ray of light or a faint spark from the induction coil had been reflected by a mirror. Not believing this possible, he passed another series of discharges through the tube, and again the same fluorescence appeared, this time looking like faint green clouds moving in unison with the fluctuating discharges of the coil. Highly excited, Röntgen lit a match and to his great surprise discovered that the source of the mysterious light was the little barium platinocyanide screen lying on the bench. He repeated the experiment again and again, each time moving the little screen farther away from the tube and each time getting the same result. There seemed to be only one explanation for the phenomenon. Evidently something emanated from the Hittorf-Crookes tube that produced an effect upon the fluorescent screen at a much greater distance than he had ever observed in his cathode ray experiments, even when he had used Lenard tubes with the thin aluminum windows.

Realizing that this conclusion was certainly in contradiction to general knowledge about cathode rays and especially his own experience that cathode rays never penetrated more than a few centimeters of air, he became deeply absorbed in attempting to explain the strange phenomenon. His concentration was so intense that he was completely unaware of the passage of time and of his surroundings. Marstaller, the diener of the institute, knocked at the door, entered the laboratory to look for a piece of apparatus, and left without being noticed. As the evening hours wore on, Röntgen's excitement in the peculiar and inexplicable observation increased. Several times Mrs. Röntgen sent a servant to call him to dinner, and when he finally sat down to the table, he ate little and in almost complete silence. The meal was hardly finished before he returned to the laboratory. He had made no reply when Mrs. Röntgen had asked him what was the matter, and his return to the laboratory in a state of suppressed excitement she ascribed to a fit of bad humor.

Although these first observations of an unidentified emanation from an excited Hittorf-Crookes tube were made on Friday, November 8, 1895, the first notes on the new phenomenon were not recorded until a few days later. Over the weekend in the absence of students there was little activity in the institute, and taking advantage of the opportunity for uninterrupted work, Röntgen returned to the laboratory early Saturday morning to repeat the experiments of the previous evening. Systematically in the next few days he made notes on the experimental set-up and his observations. Then in succeeding weeks of feverish activity he devoted himself exclusively to identifying more properties of the emanation—weeks in which he ate and even slept in his laboratory.

If the emanation could penetrate air to a hitherto unobserved degree, it was possible that it could also penetrate other substances. The inspiration came to him because of a peculiar shadow on the green fluorescent screen, apparently caused by a wire running across the tube from the induction coil. To test the truth of the conjecture he held a piece of paper, then a playing card, and then a book between the tube and the screen and closed the switch to the inductor. Simultaneously with the passage of the current through the tube the little screen behind the papers and the book lit up; the fluorescence for the book was not quite so bright as before, but it was certainly distinctly visible. He then collected some other materials, sheets of various metals, and placing them between the tube and the screen, he found that a thin aluminum sheet affected the fluorescence to approximately the same degree as had the book, but that a thin sheet of lead seemed to stop the rays completely. Already he was thinking of the new agent in terms of rays, since it had a few properties in common with known radiation, such as traveling in straight lines from the focus and throwing regular shadows. To test further the ability of lead to stop the rays, he selected a small lead piece, and in bringing it into position observed to his amazement not only that the round dark shadow of the disk appeared on the screen, but that he suddenly could distinguish the outline of his thumb and finger, within which appeared darker shadows—the bones of his hand.

One can only imagine how this first ghostly shadow picture of the human skeleton within living tissue affected the observer: Doubt must have been followed by wonder and perhaps by a reluctance to continue experiments that promised to bring him disrepute in the eyes of his colleagues. At that point he determined to continue his experiments in secrecy until such time as he himself was certain of the validity of his observations. Thus during the closing weeks of 1895 Röntgen worked in seclusion in order to prove to himself that his chance observation was a fact, and then to build up sufficient faith in his findings to hand them over to other scientists for confirmation or refutation. Even his good friend Boveri was kept in the dark, and once to Boveri's impatient question of what was going on, Röntgen replied, "I have discovered something interesting, but I do not know whether or not my observations are correct."

Knowing that cathode rays darkened a photographic plate, Röntgen set out to determine whether the peculiar radiation from the Hittorf-Crookes tube, which might or might not have been due to cathode rays, could produce a similar effect. After finding that the rays did possess this property, he extended the experiment by placing a piece of platinum on the plate before the exposure. On the developed plate a light area appeared where the platinum had absorbed the rays.

Correlating these observations with the shadow picture of the bones of his hand upon the fluorescent screen, he conceived another experiment for which one evening he persuaded Mrs. Röntgen to be the subject. At his instruction she placed her hand on a cassette loaded with a photographic plate, upon which he directed rays from his tube for fifteen minues. On the developed plate the bones of the hand appeared light within the darker shadow of the surrounding flesh; two rings on her finger had almost completely stopped the rays and were clearly visible. When he showed the picture to her, she could hardly believe that this bony hand was her own and shuddered at the thought that she was seeing her skeleton. To Mrs. Röntgen, as to many others later, this experience gave a vague premonition of death.

Once convinced that his observations were based upon

sound experimentation, Röntgen realized that early publication of his findings was essential. He spent the last days of December assembling his notes, and several days after Christmas he handed to the secretary of the Würzburg Physical Medical Society the manuscript of his paper, *On a New Kind of Rays, a Preliminary Communication,* with the unusual request that it be published in the *Sitzungsberichte* of the society, although it had not been presented at one of the meetings. The title of the paper attracted the attention of the secretary, and sensing the fundamental importance of its content, he slated it for publication in the forthcoming issue of the *Sitzungsberichte der PhysikalischMedizinischen Gesellschaft zu Würzburg,* a journal, edited by Professor Schultze, Professor Reubold, and Dr. Geigel, which was a supplement to the *Verhandlungen* of the same society.

Ueber eine neue Art von Strahlen.

von W. C. Röntgen.

(Vorläufige Mittheilung.)

1. Lässt man durch eine Hittorf'sche Vacuumröhre, oder einen genügend evacuirten Lenard'schen, Crookes'schen oder ähnlichen Apparat die Entladungen eines grösseren Ruhmkorff'schen gehen und bedeckt die Röhre mit einem ziemlich eng anliegenden Mantel aus dünnem schwarzem Carton, so sieht man in dem vollständig verdunkelten Zimmer einen in die Nähe des Apparates gebrachten, mit Bariumplatincyanür angestrichenen Papierschirm bei jeder Entladung hell aufleuchten, fluoresciren, gleichgültig ob die angestrichene oder die andere Seite des Schirmes dem Entladungsapparat zugewendet ist. Die Fluorescenz ist noch in 2 m Entfernung vom Apparat bemerkbar.

Man überzeugt sich leicht, dass die Ursache der Fluorescenz vom Innern des Entladungsapparates und von keiner anderen Stelle der Leitung ausgeht.

*On a New Kind of Rays,* first page of manuscript.

früher> Mitglieder der Gesellschaft lediglich deshalb nicht mehr im Per-
sonalverzeichnisse geführt wurden, weil sie bei ihrem Weggange aus
Würzburg vergessen hatten, den entsprechenden Antrag zu stellen.

Herr von K ö l l i k e r stellt deshalb einen Antrag auf diesbezüg-
liche Aenderung der Statuten. — Ueber denselben soll in der ersten
Sitzung des nächsten Geschäftsjahres berathen werden.

Am 28. Dezember wurde als Beitrag eingereicht:

# W. C. Röntgen: Ueber eine neue Art von Strahlen.

### (Vorläufige Mittheilung.)

1. Lässt man durch eine *Hittorf*'sche Vacuumröhre, oder
einen genügend evacuirten *Lenard*'schen, *Crookes*'schen oder ähnlichen Apparat die Entladungen eines grösseren *Ruhmkorff*'s gehen
und bedeckt die Röhre mit einem ziemlich eng anliegenden Mantel
aus dünnem, schwarzem Carton, so sieht man in dem vollständig
verdunkelten Zimmer einen in die Nähe des Apparates gebrachten,
mit Bariumplatincyanür angestrichenen Papierschirm bei jeder
Entladung hell aufleuchten, fluoresciren, gleichgültig ob die angestrichene oder die andere Seite des Schirmes dem Entladungsapparat zugewendet ist. Die Fluorescenz ist noch in 2 m Entfernung vom Apparat bemerkbar.

Man überzeugt sich leicht, dass die Ursache der Fluorescenz vom Entladungsapparat und von keiner anderen Stelle der
Leitung ausgeht.

2. Das an dieser Erscheinung zunächst Auffallende ist,
dass durch die schwarze Cartonhülse, welche keine sichtbaren
oder ultravioletten Strahlen des Sonnen- oder des elektrischen
Bogenlichtes durchlässt, ein Agens hindurchgeht, das im Stande
ist, lebhafte Fluorescenz zu erzeugen, und man wird deshalb wohl
zuerst untersuchen, ob auch andere Körper diese Eigenschaft
besitzen.

Man findet bald, dass alle Körper für dasselbe durchlässig
sind, aber in sehr verschiedenem Grade. Einige Beispiele führe
ich an. Papier ist sehr durchlässig:[1] hinter einem eingebun-

---

[1] Mit „Durchlässigkeit" eines Körpers bezeichne ich das Verhältniss der
Helligkeit eines dicht hinter dem Körper gehaltenen Fluorescenzschirmes zu der-
jenigen Helligkeit des Schirmes, welcher dieser unter denselben Verhältnissen aber
ohne Zwischenschaltung des Körpers zeigt.

*On a New Kind of Rays,* first page of printed article
in *Sitzungsberichte.*

# EINE NEUE ART

VON

# STRAHLEN.

VON

## DR. W. RÖNTGEN,

Ö. O. PROFESSOR AN DER K. UNIVERSITÄT WÜRZBURG.

WÜRZBURG.

VERLAG UND DRUCK DER STAHEL'SCHEN K. HOF- UND UNIVERSITÄTS-
BUCH- UND KUNSTHANDLUNG.
Ende 1895.

60 ₰.

*On a New Kind of Rays,* cover of first edition reprint.

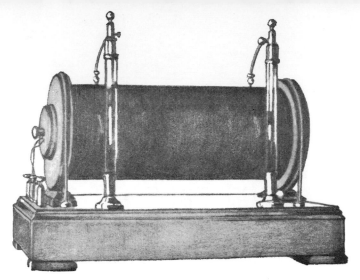

"... a fairly large Ruhmkorff induction coil ..."

"... a Hittorf vacuum tube, a sufficiently evacuated Lenard or Crookes tube, or a similar apparatus ..."

## Chapter IV *

# December 28, 1895

### W. C. Röntgen: On a New Kind of Rays
### (Preliminary Communication)

1. If one passes the discharges of a fairly large Ruhm-korff induction coil through a Hittorf vacuum tube, a sufficiently evacuated Lenard or Crookes tube, or a similar apparatus, and if one covers the tube with a rather closely fitting envelope of thin black cardboard, one observes in the completely darkened room that a piece of paper painted with barium platinocyanide lying near the apparatus glows brightly or becomes fluorescent with each discharge, regardless of whether the coated surface or the other side faces the discharge apparatus. The fluorescence is still visible at a distance of 2 m from the apparatus.

One easily convinces oneself that the cause for the fluorescence emanates from the discharge apparatus and from no other point in the circuit.

2. Observing this phenomenon one is immediately struck by the fact that the black cardboard cover, which stops visible or ultraviolet rays from the sun or the electric arc, transmits an agent that can produce active fluorescence, and one would therefore wish to investigate first whether other materials also possess this same property.

One soon finds that all materials are transparent to it, although differing widely in degree. I present a few ex-

* This new translation of Röntgen's first communication varies somewhat from one made by G. F. Barker in 1896.

amples. Paper is very transparent[1]: I observed that the fluorescent screen still glowed brightly behind a bound book of about 1000 pages; the printer's ink had no noticeable effect. Likewise fluorescence appeared behind a double pack of Whist cards; the eye can hardly detect a single card held between the apparatus and the screen.—Also a single sheet of tinfoil is hardly perceptible; only after several layers have been placed one on top of the other does one see the shadow distinctly on the screen.—Thick blocks of wood are also very transparent; pine boards 2 to 3 cm thick absorb only very little.—A plate of aluminum about 15 mm thick reduced the effect considerably but did not make the fluorescence disappear entirely.—Sheets of hard rubber several centimeters thick also let rays pass through.[2] Glass plates of equal thickness act differently depending upon whether or not they contain lead (flint glass); the former are much less transparent than the latter.—If one holds the hand between the discharge apparatus and the screen, one sees the darker shadows of the bones within the much fainter shadow picture of the hand itself.—Water, carbon disulfide, and several other liquids, when examined in mica containers, were found to be very transparent.—I have not been able to determine that hydrogen is definitely more transparent than air.—Fluorescence still may be clearly detected behind plates of copper, silver, lead, gold, or platinum, but only if the plates are not too thick. Platinum 0.2 mm thick is also transparent; silver and copper plates

[1] "Transparency" of a material I define as the ratio of the brightness of a fluorescent screen placed directly behind the material to the brightness of the screen under identical conditions without interposition of the material.

[2] For the sake of brevity I should like to use the term "rays," and to distinguish them from others I shall use the name "x-rays." (See no. 14.)

may even be thicker. Lead 1.5 mm thick is practically opaque and on account of this property was frequently used. —A stick of wood having a square cross section (20 by 20 mm) and one side painted white with lead paint acts differently depending upon how it is held between apparatus and screen; although there is practically no effect if the direction of the x-rays is parallel to the painted surface, the stick throws a dark shadow if the rays have to pass through the painted surface.—In a manner similar to that of the metals themselves, their salts, either solid or in solution, may be arranged according to their transparency.

3. The experimental results cited, as well as others, lead to the conclusion that the transparency of various substances assumed to be of equal thickness depends primarily upon their density: no other property, at least not to the same extent, is as conspicuous as this one.

That the density, however, is not the only determining factor is proved by the following experiments. I studied the transparency of plates of almost equal thickness made of glass, aluminum, calcite, and quartz; although the density of these substances is approximately the same, it was quite evident that calcite was considerably less transparent than the other materials, which all reacted very much the same. I have not noticed a particularly strong fluorescence of calcite, especially as compared with glass (see no. 6).

4. As the thickness increases all materials become less transparent. In order to find a possible relation between transparency and thickness, I made photographs (see no. 6) in which the photographic plate was partly covered with a number of layers of tinfoil in a steplike arrangement; a photometric measurement will be made when I have a suitable photometer.

5. Platinum, lead, zinc, and aluminum were rolled out in sheets of such thickness that all appeared nearly equally transparent. The following table contains the measured thickness in millimeters, the relative thickness referred to that of the platinum sheet, and the density.

|    | Thickness |    | Relative Thickness | Density |
|----|-----------|-----|--------------------|---------|
| Pt | 0.018     | mm | 1                  | 21.5    |
| Pb | 0.05      | mm | 3                  | 11.3    |
| Zn | 0.10      | mm | 6                  | 7.1     |
| Al | 3.5       | mm | 200                | 2.6     |

These values show that by no means is the transparency of different metals equal if the product of thickness and density is the same. The transparency increases much more rapidly than the product decreases.

6. The fluorescence of barium platinocyanide is not the only detectable effect of x-rays. First it must be mentioned that other substances also fluoresce, such as, for example, the phosphorescent calcium compounds, uranium glass, ordinary glass, calcite, rock salt, and so forth.

Of special significance in many respects is the fact that photographic dry plates are sensitive to x-rays. One is able to make a permanent record of many phenomena whereby deceptions are more easily avoided; and as a control I have, whenever possible, recorded every relatively important observation that I saw on the fluorescent screen by means of photography.

Here, the property of the rays of penetrating almost unhindered thinner layers of wood, paper, and tinfoil is very advantageous; in the lighted room one can expose the photographic plate, which is enclosed in a cassette or wrapped in paper. On the other hand, as a consequence of this property one should not leave undeveloped plates near the discharge apparatus for any length of time if these plates are protected merely by the ordinary cardboard box and paper.

The question remains whether or not x-rays are directly responsible for the chemical action upon the silver salts of the photographic plate. It is possible that this action is due to the fluorescent light which, as indicated above, is produced in the glass plate or perhaps in the gelatin layer. "Films," by the way, may be used equally as well as glass plates.

That x-rays are also able to produce a heating action I have not yet proved experimentally; yet one may well assume that this effect exists, since the fluorescent phenomena prove that x-rays may be transformed and since it is also evident that not all of the impinging x-rays leave the material unaltered.

The retina of the eye is insensitive to our rays; the eye brought close to the discharge apparatus registers nothing, although, according to common experience, the media contained in the eye must be sufficiently transparent to the rays.

7. After I had recognized the transparency of various relatively thick materials, I was anxious to learn how x-rays behaved when passing through a prism, that is, whether or not they were refracted by it. Experiments with water and with carbon disulfide in mica prisms having a refracting angle of approximately 30 degrees did not show any refraction either on the fluorescent screen or on the photographic plate. As a control the refraction of light rays was observed under the same conditions; the refracted images on the plate were found to be located about 10 and 20 mm respectively from the nonrefracted.—With hard rubber and aluminum prisms, also of a refracting angle of about 30 degrees, I have obtained images on the photographic

plate in which one might possibly detect a refraction. However, this is very uncertain, and if refraction does exist, it is in any case so small that the refractive index of the x-rays in these substances could not be more than 1.05 at the most. Also on the fluorescent screen I was unable to observe any refraction in this instance.

Experiments with prisms of denser metals have not as yet produced any definite results owing to their low transparency and the resultant low intensity of the transmitted rays.

Considering these facts on one hand and on the other the importance of the question whether or not x-rays can be refracted when passing from one medium into another, it is reassuring that this question can be investigated in a different manner without the aid of prisms. Finely pulverized substances in sufficiently heavy layers scatter the impinging light and because of refraction and reflection let pass only a small amount of it; now, if the powders are equally as transparent to x-rays as the coherent substance is—provided that equal masses of each are used—, it follows that neither refraction nor regular reflection takes place to any appreciable degree. Such experiments were carried out with finely pulverized rock salt, with fine silver powder produced electrolytically, and with zinc dust such as is frequently used in chemical investigations; in all cases no difference in the transparency between powder and coherent substance could be detected, neither with the fluorescent screen nor with the photographic plate.

That one cannot concentrate x-rays with lenses is self-evident from the foregoing; indeed a large hard rubber lens and a glass lens were ineffective. The shadow picture of a round rod is darker in the center than at the edge; that of a tube, filled with a substance more transparent than the material of the tube itself, is lighter in the center than at the edge.

8. On the basis of the preceding paragraph the question in regard to the reflection of x-rays may be considered solved in the sense that a noticeable regular reflection of the rays from any of the examined substances did not take place. Other experiments, which I shall omit here, lead to the same result.

However, one observation must be mentioned, which at first seems to be contradictory. I exposed to x-rays a photographic plate that was protected from light by black paper with the glass side pointing toward the discharge apparatus; the sensitive layer with the exception of a small free space was covered with polished sheets of platinum, lead, zinc, and aluminum in a starlike arrangement. On the developed negative one can clearly perceive that the blackening under the platinum, the lead, and particularly under the zinc is denser than under the other areas; the aluminum had exerted no effect. It seems, therefore, that the three named metals reflect the rays; however, one may conceive of other causes for the stronger blackening, and in the second experiment, in order to be sure, I placed a piece of thin aluminum foil, which is not transparent to ultraviolet rays but is very transparent to x-rays, between the sensitive layer and the metal plates. Since essentially the same result was again obtained, a reflection of x-rays from the aforementioned metals is proved.

If one adds to this fact the observation that powders are equally as transparent as coherent materials and furthermore that materials having rough surfaces have the same effect upon the transmission of x-rays as polished substances as described in the last experiment, one comes to the conclusion that, although, as stated before, a regular reflection does not take place, materials react to x-rays as do turbid media to light.

Since, moreover, I could not detect any refraction when x-rays pass from one medium into another, it appears that they move with equal velocity in all materials, specifically in a medium that is present everywhere and in which particles of matter are embedded. These particles form an obstacle to the propagation of x-rays, which in general is the greater the denser the respective substance.

9. Therefore the arrangement of particles within the material may possibly influence its transparency; for instance, a piece of calcite of a given thickness may vary in transparency depending upon whether the rays pass through it in the direction of its axis or at right angles thereto. Experiments with calcite and quartz, however, have given a negative result.

10. It is well known that Lenard, in his beautiful experiments on Hittorf's cathode rays passing through a thin aluminum foil, came to the conclusion that these rays are phenomena in the ether and that they are diffused in all materials. Regarding our rays we can make similar statements.

In his recent publication Lenard determined the absorption of cathode rays in different materials, and, among others, for air of atmospheric pressure he found it to be 4.10, 3.40, and 3.10, all relative to 1 cm, depending upon the rarefaction of the gas in the discharge apparatus. In my experiments, judging from the discharge voltage estimated from the spark gap, I was dealing usually with rarefactions of approximately the same order of magnitude and only occasionally with higher or lower ones. With L. Weber's photometer—I do not have a better one—I succeeded in comparing in atmospheric air the intensities of the fluo-

rescent light of my screen at two distances from the discharge
apparatus—about 100 and 200 mm respectively—and I
found in three experiments, which were in very good agreement, that they were inversely proportional to the squares
of their respective distances between screen and discharge
apparatus. Therefore air absorbs a much smaller portion of
transmitted x-rays than of cathode rays. This result is also in
entire agreement with the previously mentioned observation
that the fluorescent light may still be observed at a distance
of 2 m from the discharge apparatus.

In general other substances have properties similar to
air: They are more transparent to x-rays than to cathode
rays.

11. Another very remarkable difference between the behavior of cathode rays and of x-rays lies in the fact that,
despite many attempts, I have not succeeded in obtaining
a deflection of the x-rays by a magnet, even in very strong
magnetic fields.

So far, the deflection by means of a magnet has been
considered a property peculiarly characteristic of cathode
rays; it is true that Hertz and Lenard observed that there
are different kinds of cathode rays "which can be differentiated from one another by their production of phosphor-
escence, by their absorption, and by their deflection by a
magnet," but a considerable deflection was found in all
their investigations, and I do not believe that one should
give up this characteristic feature without good reason.

12. According to experiments made especially for this
purpose it is certain that the area on the wall of the discharge apparatus that shows the strongest fluorescence must
be considered the main point of emission of x-rays, which
radiate in all directions. Therefore, the x-rays proceed from
that area where, according to the reports of several inves-

tigators, the cathode rays impinge upon the glass wall. If one deflects the cathode rays within the discharge apparatus by means of a magnet, one observes also that the x-rays are emitted now from another area, namely, from the terminating point of the cathode rays.

This is another reason why x-rays, which cannot be deflected, cannot be simply cathode rays that have been transmitted or reflected without being changed by the glass wall. The greater density of the glass outside the discharge tube cannot, according to Lenard, be made responsible for the great difference in deflection.

I therefore come to the conclusion that x-rays are not identical with cathode rays, but that they are produced by the cathode rays in the glass wall of the discharge apparatus.

13. This production takes place not only in glass but also in aluminum, as I was able to observe with an apparatus sealed with an aluminum window 2 mm thick. Other substances are to be examined later.

14. I find the justification for using the name "rays" for the agent emanating from the wall of the discharge apparatus in the very regular formation of shadows that are produced if one brings more or less transparent materials between the apparatus and the fluorescent screen (or the photographic plate).

I have observed and sometimes photographed many such shadow pictures, the production of which is occasionally very attractive; for instance, I have photographs of the shadows of the profile of a door separating rooms, in one of which the discharge apparatus was placed and in the other the photographic plate; of the shadows of the bones of the hand; of the shadows of a concealed wire wound on a wooden spool; of a set of weights enclosed in a small box; of a compass in which the magnetic needle is entirely sur-

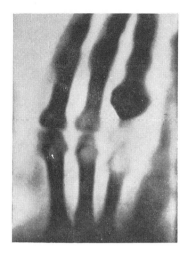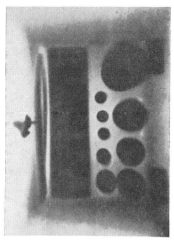

". . . shadows of the bones of the hand; of a set of weights enclosed in a small box . . ."

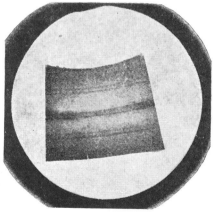

". . . shadows of a compass in which the magnetic needle is entirely surrounded by metal; of a piece of metal whose inhomogeneity becomes apparent with x-rays . . ."

rounded by metal; of a piece of metal whose inhomogeneity becomes apparent with x-rays; and so forth.

That x-rays are propagated in straight lines is further proved by a pinhole photograph that I was able to make of the discharge apparatus enclosed in black paper; the picture is weak but unmistakably correct.

15. I have often looked for interference phenomena of x-rays but unfortunately without success, possibly only because of their low intensity.

16. Experiments to determine whether or not electrostatic forces can affect x-rays have been started but as yet are not finished.

17. If one asks oneself what x-rays—which cannot be cathode rays—really are, at first, misguided by their lively fluorescence and chemical effects, one might perhaps think of ultraviolet light. However, one is immediately confronted with rather serious considerations. For, if x-rays were ultraviolet light, this light should have the following properties:

(a) that, in passing from air into water, carbon disulfide, aluminum, rock salt, glass, zinc, and so forth, it suffers no noticeable refraction;

(b) that it cannot be regularly reflected to any noticeable extent by these substances;

(c) that, therefore, it cannot be polarized by any of the ordinary methods;

(d) that no other property of the material influences its absorption as much as its density.

In other words, one would have to assume that these ultraviolet rays behave entirely differently from the infrared, visible, and ultraviolet rays known at present.

I have not been able to arrive at this conclusion and have sought for another explanation.

Some kind of relation seems to exist between the new rays and light rays, at least as is indicated by the formation of shadows, by fluorescence, and by chemical effects, which are common to both types of rays. Now it has been known for a long time that, besides transversal light vibrations, longitudinal vibrations in the ether can also occur and must even exist according to the opinion of several physicists. It is true that their existence has not yet been definitely proved and that therefore their properties have not yet been investigated experimentally.

Could not, therefore, the new rays be due to longitudinal vibrations in the ether.

I must confess that during the course of investigations I have favored this thought more and more, and I therefore take the liberty of expressing this theory here, although I am perfectly aware that the explanation offered requires further confirmation.

Würzburg. Physikal. Institut der Universität, Dec. 1895.

## Chapter V
# January—March 1896

ETWEEN THE SUBMITTING of the manuscript on *A New Kind of Rays* and its appearance in print, Röntgen knew uncertainty. He was aware that the evidence he had accumulated by careful experimentation might not be accepted or, if accepted, might not be substantiated. The desire that investigations of the properties of the new rays should be continued was offset by the fear that his report might be consigned to oblivion with other observations of interesting but unorthodox physical phenomena.

His working habits and a natural reserve denied Röntgen the reassurance he might have had if he had confided in his assistants and had received from them a foretaste of either enthusiasm or criticism. Close as Mrs. Röntgen was to him, particularly during this important period of experimentation, and genuine as was her interest, she was hardly equipped to evaluate his researches accurately. So it was that Röntgen weathered the uncertainty more or less alone.

The communication appeared on pages 132 to 141, the last pages, of the 1895 volume of the *Sitzungsberichte* of the society, published and printed by the Stahel'sche K. Hof- und Universitäts-Buch- und Kunsthandlung in Würzburg. Reprints were bound in yellow wrappers and carried the title page inscription *"Eine neue Art von Strahlen*—von Dr. W. Röntgen, ö. o. Professor an der K. Universität Würzburg.—Ende 1895—60 Pfennig." Although Röntgen had as usual included his middle initial *C*. in his signature on the manuscript, the reprint of the first communication was marked by its omission.

To speed critical reading and evaluation of the work, he chose recognized physicists, many of whom he knew as friends, and to them on New Year's Day in 1896, a Wednesday, he addressed copies of the reprint, with some enclosing prints of the x-ray pictures he had taken. Among these physicists were E. Warburg and Lummer in Berlin, F. Exner in Vienna, Voller in Hamburg, Kohlrausch in Strassburg, Zehnder in Freiburg i. Br., Schuster in Manchester, and Poincaré in Paris. "Now the devil will be to pay," he said to his wife after dropping the reprints in the big yellow box on the Pleicher Ring.

In Würzburg only a few vague rumors were being circulated. At the institute speculation was rife, and under the seal of absolute secrecy members of the personnel told impressive stories about the mysterious rays. Those who knew the most, which was very little—Siebenlist, the photographer who had helped Röntgen with the developing of plates exposed to x-rays and the printing of photographs, and colleagues at the university who had supplied him with lead, zinc, aluminum, and tinfoil—remained discreet.

Then the day came early in the new year when Röntgen was no longer a middle-aged professor of physics at the University of Würzburg but the focus of international praise, condemnation, and curiosity. From all over the world came letters of congratulation and incredulity as well as reports of duplication of the original experiments and a few of failures. To the Würzburg institute were sent tubes of various construction and other equipment for the production of the rays. And through its modest doors passed scientists, reporters, the sympathetic, and the curious. The location of the Röntgen residence in the institute left no escape from the heterogeneous horde that descended upon it. "Our domestic peace is gone," complained Mrs. Röntgen to a friend, and Röntgen was forced to adjust from the satisfying freedoms of a quiet private life to a tacit acceptance of public demands. Some visitors went so far as to filch x-ray photographs from the laboratory, and post cards with Röntgen's signature failed to reach their destination.

The response of scientists and laymen seemed to him decidedly out of proportion to the simple, unpretentious, rather dry style of the published communication. Unquestionably except for the many pictures of hands made quickly after the communication was published, which demonstrated the importance of the new rays in the study of anatomic structures and pathologic changes, the discovery might have been consigned for some time to the relative oblivion of the physical laboratory.

Among the many early expressions of recognition that had real meaning for Röntgen was one from Warburg in Berlin received on Saturday, January 4. In addition to acknowledging the reprint and the x-ray pictures, Warburg wrote that he had taken the liberty of adding the photographs to a temporary exhibit at the physical institute of Berlin University, on the occasion of the fiftieth anniversary of the Berlin Physical Society. The transactions of the celebration mailed to Röntgen a few days later carried an announcement: "A series of photographs is on exhibit which Herr Röntgen of Würzburg has taken with the x-rays recently discovered by him." In a footnote von Bezold, the president of the Berlin Society, referred to this display of Röntgen's pictures:

Unfortunately the speaker and many other members of the society were so busy that evening that they did not know that among the exhibits were Röntgen's first photographs in a secluded corner hidden by many spectators. If the speaker had heard but one word about these photographs, he would have closed his speech at the banquet quite differently and would have called attention to the rare inspiration that was given the occasion by a preliminary communication on a discovery so important that its significance could not fail to be appreciated, even at first glance.

In contrast sensational stories about the photographs with a flavor not much to Röntgen's liking appeared in the second morning edition of the *Frankfurter Zeitung* on Tuesday, January 7. The popularization of science in lectures or in print, Röntgen always claimed, did much more harm than good.

A Sensational Discovery—In scientific circles of Vienna the news of a discovery, made by the professor of physics Wilhelm Conrad Röntgen of Würzburg, is being discussed enthusiastically. If this discovery fulfils its promise, it constitutes an epoch-making result of research in exact science, which is destined to have interesting consequences along medical as well as physical lines. The Vienna *Presse* reports as follows: "Professor Röntgen takes a Crookes tube—a strongly evacuated glass tube through which an induction current is passed—and takes photographs by means of rays, which are emitted by this tube into space, using ordinary photographic plates. These rays, the existence of which so far has been unknown, are entirely invisible to the eye; contrary to ordinary rays they penetrate wood, organic and other nontransparent materials. Metals and bones, however, stop the rays. One can photograph in plain daylight with a 'closed cassette.' This means not only that the light rays follow the ordinary path but also that they penetrate the wooden cover, which is placed in front of the light-sensitive plates and which ordinarily must be removed before a photograph is taken. They also penetrate a wooden cover in front of the object to be photographed. Professòr Röntgen, for instance, took a photograph of a set of weights without opening the wooden box in which the weights were kept.

A few examples of this sensational discovery are being circulated in scientific circles in Vienna and deservedly are creating great amazement. The matter will be tested very carefully in the near future in the laboratories and probably will be developed further. The physicists must study this unknown radiation, which is capable of penetrating matter ordinarily opaque to light. The light rays from a Crookes tube penetrate dense objects as easily as sunlight penetrates a piece of glass. Biologists and physicians, especially surgeons, will be very much interested in the practical use of these rays, because they offer prospects of constituting a new and very valuable aid in diagnosis.

At the present time, we wish only to call attention to the importance this discovery would have in the diagnosis of diseases and injuries of bones, provided that the process can be developed technically so that not only the human hand can be photographed but that details of other bones may be shown without the flesh. The surgeon then could determine the extent of a complicated bone fracture without the manual examination which is so painful to the patient; he could find the position of a foreign body, such as a bullet or piece of shell, much more easily than has been possible heretofore and without any painful examinations with a probe. Such photographs also would be extremely valuable in diagnosing bone diseases which do not originate from an injury and would help to guide the way in therapy.

And in another edition of the *Frankfurter Zeitung* on the following evening:

There are nine photographs in Vienna which Professor Röntgen has sent to one of his colleagues. The most careful examinations of these photographs leave no doubt concerning the complete validity of Professor Röntgen's statements. The more strictly and carefully one examines them, the more convincing these pictures become. The Würzburg professor discovered these unknown rays by accident, as happens so often when sensational truths are disclosed. He had covered a Crookes tube with cloth, and, in making a certain experiment, he sent a strong electric current through this tube, which he had placed on his laboratory table. He noticed that a piece of sensitized paper, which was lying on the table, showed certain lines that were not there before. The perspicacious professor followed up this observation, and the familiar results of his discovery have just been reported.

It was not odd, although it had not occurred to Röntgen, that the use of the rays in photography had been largely responsible for the immediate and excited public response. That they thus might have use in medical diagnosis was another factor in the discovery with decided public appeal. So a newspaper storm broke, which to Röntgen was extremely distasteful. In the pattern of the story in the Vienna *Presse* the London *Standard*, the Paris *Matin*, and other papers in all parts of the world gave space to the discovery of the rays, embellishing it often with factual but more often with fantastic speculations of their own and frequently misspelling the discoverer's name "Routgen" or "Rothgen."

Years later the reason for the undigested and premature news dispatches emanating from Vienna was revealed. Among the reprints mailed to his fellow physicists on New Year's Day Röntgen addressed one, including prints of his photographs, to his old friend Franz Exner, with whom he had worked in his youth in Kundt's laboratory and who was now a professor of physics in Vienna. With a well founded enthusiasm in Röntgen's discovery Exner at a party in his home showed the photographs to friends and subsequently lent them to his colleague Ernst Lecher, who had come over from Prag. Lecher in turn showed them at once to Z. K. Lecher, his father. The elder Lecher, who was then editor of the old Vienna *Presse*, lost no time in exploiting the enormous news value in the story of the rays' discovery. His was the original contention that the discovery might become of extreme importance in the study of disease, and his enthusiastic article appeared on the front page of the Sunday, January 5, 1896, edition of the *Presse*. It became the prototype for the first stories to appear on the discovery. The Vienna representative of the London *Daily Chronicle* reported the news at once to his home office, and from London on Monday, January 6, 1896, the discovery was disseminated by cable in the following words:

The noise of the war's alarm should not distract attention from the marvelous triumph of science which is reported from Vienna. It is announced that Prof. Routgen of the Wurzburg University has discovered a light which for the purpose of photography will penetrate wood, flesh, cloth, and most other organic substances. The professor has succeeded in photographing metal weights which were in a closed wooden case, also a man's hand which showed only the bones, the flesh being invisible.

On January 9 Röntgen's hometown paper, the *Würzburger Generalanzeiger*, belatedly boarded the bandwagon and largely repeating the other news reports introduced them in the following vein:

On a New Kind of Rays—Last month Dr. W. C. Röntgen, Professor at the University, gave a lecture before the Würzburg Physical-Medical Society on a discovery that he made and that is termed epoch-making and sensational in long articles published in scientific publications. . . .

Assuming that the scientific societies met during the Christmas holidays, the Würzburg paper was in error in stating that Röntgen had first presented his discovery in a lecture.

The newspaper speculations on the medical use of x-ray photographs, although seemingly unwarranted at first, were quickly picked up and shared by serious scientists and practical physicians, who offered incontestable proof of their soundness. The *New York Medical Record*, the *Lancet*, and the *British Medical Journal* for January 11, the *Münchner Medizinische Wochenschrift* for January 14, the *Wiener Klinische Wochenschrift* for January 16, the *Comptes Rendus* for January 20, the *Settimana Medica* for January 25, and the *Journal of the American Medical Association* for February 15, among others, printed articles on the value of x-rays in medicine.

Articles also appeared in nonmedical scientific journals, in the *New York Electrical Engineer* for January 8, the *London Electrician* for January 10, *Nature* for January 16, *Il Nuovo Cimento* for January, and *L'Eclairage Electrique* for February 8.

More and more letters continued to pour in, some containing complimentary messages, some reflecting envy, and a few expressing fear of the "death's rays." Among the letters addressed to Röntgen containing appeals for help was one from a locksmith, whose young son some time before had broken his leg. The leg had healed but continued to get shorter and shorter in comparison with the other. A physician had x-rayed the leg and had told the father that the bones would have to be broken again in order to fit the ends together properly. Wouldn't it be advisable, the father wrote, as long as one leg had to be broken, to break the bones of the other leg in such a way that it would be shortened by the same length. "A rather workman-like idea of the locksmith," Röntgen said. "A physician would hardly have thought of that."

Another incident took place, which Röntgen often told with great amusement. "In Vienna a public lecture on x-rays was planned, and the police department had been asked for permission; it decided: 'The experiment with x-rays cannot take place until further notice, because no details about it have been learned by this office.' And this in March 1896."

Röntgen made no attempt to answer any of the communications, but one Saturday night he sat down and aired his views on the commotion attendant to his discovery to his old co-worker and friend Zehnder:

Good friends come last; that's the way it goes. But you are the first to receive an answer. Many thanks for everything you wrote me. I cannot as yet make use of your speculations on the nature of the x-rays, since it does not seem permissible or propitious to attempt to explain a phenomenon of unknown nature with a, to me, not entirely unobjectionable hypothesis. Of what nature the rays are, is not altogether clear to me; and whether they are actually longitudinal light rays is to me of secondary importance. The facts are the important thing. In this respect my work has received recognition from many quarters. Boltzmann, Warburg, Kohlrausch, and (last but not least) Lord Kelvin, Stokes, Poincaré, and others have expressed to me their joy over the discovery and their appreciation. That is worth a great deal to me, and I let the envious chatter in peace; I am not concerned about that.

I had not spoken to anyone about my work; to my wife I mentioned merely that I was doing something of which people, when they found out about it, would say: "Röntgen seems to have gone crazy." On the first of January I mailed the reprints, and then hell broke loose! The Vienna *Presse* was the first to blow the trumpet of advertising, and the others followed. In a few days I was disgusted with the business; I could not recognize my own work in the reports anymore. For me photography was the means to the end, but they made it the most important thing. Gradually I became accustomed to the uproar, but the storm cost time; for exactly four weeks I was unable to make a single experiment. Other persons could work, only I could not. You have no conception how upset things were here.

I am enclosing the promised photographs; if you wish to show them in lectures, it is all right with me; but I would suggest that you put them under glass and frame. Otherwise they will be stolen. I think that with the aid of the legends you will have no difficulty with them; if you do, write me.

I use a large Ruhmkorff 50/20 cm with a Deprez inter-rupter, and about 20 ampères primary current. My apparatus, which remains on the Raps pump, requires several days for evacuation; the best results are obtained when the spark gap, connected in parallel, is about 3 cm.

In time all discharge apparatus will be punctured. Any method of producing cathode rays will be successful, also with incandescent lamps according to Tesla and with tubes without electrodes. For photography I use three to ten minutes depend-ing upon the conditions of the experiment—Best regards from house to house.

While many other scientists were working feverishly to discover new properties and uses for the rays, Röntgen, who was most eager to follow up his own experiments, had great difficulty escaping the numerous demands made upon him. One reporter, who gave him credit for having made the greatest discovery in a century and who suggested that he should have his portrait painted with his tubes, Röntgen politely dismissed with, "There is much to do and I am busy, very busy." To him it was inconceivable that so much glory should be showered upon him.

It was only a short time after the x-rays were publicized that Röntgen was summoned to appear at the Im-
perial Court in Berlin to demonstrate his discovery to the
Emperor, Wilhelm II. Protected from the bitter winter
weather by a bulky fur-lined coat and carrying a knobby
brief case, he arrived at the royal castle late on the afternoon
of January 13, 1896 with these brief and unorthodox words
of apology: "I beg pardon, your Majesty, for being late,
but I am not used to these large distances here in Berlin."

He set up his apparatus in the Sternensaal of the castle,
and then before the Emperor, the Empress Augusta Victoria, the ex-Empress Friederich, and many distinguished
guests, including General Leuthold, the chief army physician, he demonstrated with a fluorescent screen how x-rays
penetrated wooden boards and cardboard boxes. He also
photographed a few lifeless objects. Since the tube was the
only one he had left at his disposal and would not lend
itself to further strain, he did not attempt a picture of the
human hand. "I hope I shall have 'Kaiser luck' with this
tube," Röntgen had said before the demonstration, "for
these tubes are very sensitive and are often destroyed in
the very first experiment, and it takes about four days to
evacuate a new one."

The x-ray photographs he had brought with him created
quite a sensation. In the discussion after the demonstration
it was even suggested that x-rays might throw light on the
secrets of gravitation or perhaps make practical use of that
force. In recognition of his distinguished contribution to
science, Röntgen was decorated by the Emperor with the
Prussian Order of the Crown, II Class. Later, seated at
dinner between the Emperor and Count Moltke, Röntgen
found himself the target for more questions about future
developments of the use of the rays, and it was after mid-

night when he left the castle. Having honest pride in the honor his discovery had brought him and feeling again the need to prove himself worthy, he resolved henceforth to intensify his study of the rays.

Among many invitations to explain his findings in an open lecture was one from the German parliament, the Reichstag. This he felt compelled to refuse, as well as those from numerous scientific societies, in order that he might continue his own experiments. But he could not ignore an urgent appeal for a lecture from his Würzburg colleagues, reinforced by the voice of K. B. Lehmann, president of the Würzburg Physical-Medical Society and professor of hygiene at the university.

It was to be expected that the first lecture in Röntgen's own institute, scheduled for Thursday night, January 23, 1896, would attract many members of the nobility, university professors, high city officials, high ranking army officers, and students. Every seat in the auditorium was filled long before the meeting began, and upon Röntgen's entrance a veritable storm of applause broke, which was recurrent many times during the evening. Röntgen began to speak modestly about his work. First he emphasized that, because of the general interest aroused, he considered it his duty to speak publicly about his *Arbeit,* even though the experiments were still in the preliminary stage. He then mentioned the work of Hertz, Lenard, and others and said that his own observations had led him to make experiments along the same lines. He described the fluorescence of his little barium platinocyanide screen and told how he had quickly found that the tube itself was responsible for the strange phenomenon. "I found by accident," he said, "that the rays also penetrated black paper. I then used wood, paper, books, but I still believed I was the victim of deception. Finally I used photography, and the experiment was successfully culminated."

Röntgen then demonstrated the power of the rays to penetrate paper, tin, wood, and his own hand and showed that a platinum foil stopped them. He reported his early attempts to make x-ray photographs through a door in his laboratory, separating the room in which the induction coil and discharge tube were located and the room containing the photographic plate. After the developing, the plate had shown several light strips, which at first he had been unable to explain. He had had the door dismantled and reported, "The different shadows on the plate showed me that they were not caused by the various thicknesses but by a surface absorption of the door. I found out that the door was covered by white lead, and since lead absorbs these rays considerably, it is easy to see that a lead layer running in the direction of the rays absorbs more than one through which the rays go perpendicularly." Röntgen then showed various x-ray pictures, of weights in a box, a compass, a wire wound on a piece of wood, and finally the picture of a human hand. These pictures also excited extreme interest and brought forth favorable comment.

Röntgen then asked His Excellency Albert von Kölliker, the famous anatomist of the university, for permission to photograph his hand. Von Kölliker eagerly complied, and a little later the excellent x-ray picture of this hand was shown to the audience amid tremendous applause. Von Kölliker said with feeling that during his forty-eight years as a member of the Physical-Medical Society he had never attended a meeting with a presentation of greater significance, neither in the field of natural sciences nor probably in medical science. After leading the audience in three cheers for the discoverer, he proposed that henceforth the rays be called "Röntgen's rays," and again the crowd expressed approval.

In a short discussion after the meeting, carried on later at length over a glass of beer, von Kölliker wished to know whether it would be possible to make x-ray photographs of other parts of the human body, and whether possibly surgery and anatomy might not benefit by the discovery. According to Röntgen's demonstration an apparent obstacle to this further development was the approximately equal density of the different organs, nerves, muscles, and veins, which hence, unlike the bones, probably could not be differentiated by the rays, which only produced a definite shadow of bones. Schönborn, a surgeon, warned against too much optimism and doubted that the method would ever be of appreciable value in the diagnosis of internal conditions. Röntgen replied that according to his method it was not difficult to photograph a dog or a cat and that it should be possible soon to make x-ray pictures of larger parts of the human body. Without having time himself to continue experiments in that direction, he expressed a willingness to give the benefit of his experience to anyone undertaking such experiments in a medical institution.

This memorable lecture probably was the only one on the discovery that Röntgen ever gave before a large audience.

The hectic course of events was naturally reflected in the Röntgen household and caused a minimum of distress only because of Mrs. Röntgen's skilful handling of the situation. On March 4, 1896 she wrote to her husband's cousin Louise, who had meanwhile married an Indianapolis minister, J. G. Grauel.

Wilhelm has so much work he doesn't know which way to turn. Yes, dear Louise, it is no small matter to become a famous man, and few persons realize how much work and unrest this carries with it. . . . When Wilhelm told me in November that he was working on an interesting problem, we had no idea how it would be received, but as soon as the paper was published our domestic peace was gone.

Every day I am astonished at the enormous working capacity of my husband, and that he can keep his thoughts on his work in spite of the thousand little things with which he is annoyed. But now it is high time that he should rest, and I am preparing everything for our departure. We are going south for a few weeks, in order to permit Wilhelm to spend all of his time in the open. Every day I am grateful to God that He made him so healthy and strong, and yet often I am fearful that some day the strain may become too much for him.

But now I talk only of the less lovely part of our experience and have not said a word about our great happiness over the success of his work. Our hearts are full of gratitude that we are permitted to live through such a wonderful experience. How many recognitions has my dear received for his indefatigable research. Often we are almost dizzy with all the praise and honors. It would be alarming if the man who received all this were vain. But you know my honest modest husband as scarcely anyone else does, and you can understand that he finds his highest award in the fact that he was permitted to accomplish something valuable in serving pure science. . . .

At about this time Röntgen prepared the text of his second communication. This he placed before the Redaktions Commission of the *Sitzungsberichte* of the Physical-Medical Society on March 9, 1896.

## Chapter VI

# March 9, 1896

## W. C. Röntgen: On a New Kind of Rays
### *(Continued)*

Since my work must be interrupted for several weeks, I should like to present at this time some new results in the following.

18. At the time of my first publication I knew that x-rays are able to discharge electrified bodies, and I suspect that in Lenard's experiments it was also the x-rays and not the cathode rays, transmitted unchanged by the aluminum window of his apparatus, that produced the effects upon electrified bodies at a distance. However, I waited until I could present incontestable results before publishing my experiments.

These seem to be obtainable only if the observations are made in a room that not only is protected completely from the electrostatic forces emanating from the vacuum tube, from the conducting wires, from the induction apparatus, and so forth, but also is closed against air that comes from the region of the discharge apparatus.

Accordingly I had a box built of zinc plates soldered together, which is large enough to accommodate me and the necessary instruments and which is completely airtight with the exception of an opening that could be closed by a zinc door. The wall opposite the door is to a large extent covered with lead; at a place near the discharge apparatus, which is set up outside the box, an opening 4 cm

---

* This new translation of Röntgen's second communication varies somewhat from one made by G. F. Barker in 1896.

wide is cut out of the zinc wall and its lead cover, and this opening is in turn made air-tight with a thin sheet of aluminum. Through this window the x-rays can enter the observation box.

Now I observed the following:

(a) Positively or negatively electrified bodies set up in air are discharged if they are irradiated with x-rays; the more intense the rays, the more rapid the discharge. The intensity of the rays was estimated by their effect upon the fluorescent screen or upon a photographic plate.

Generally it is immaterial whether the electrified bodies are conductors or insulators. Moreover, so far I have not been able to find a specific difference in the behavior of different bodies with regard to the rate of discharge, nor in the behavior of positive and negative electricity. Yet it is not impossible that small differences exist.

(b) If an electrified conductor is not surrounded by air but by a solid insulator, e.g., paraffin, the irradiation of it has the same effect as moving a grounded flame over the insulating cover.

(c) If this insulating cover is surrounded by a tight-fitting grounded conductor, which like the insulator must be transparent to x-rays, the radiation exerts upon the inner electrified conductor no effect detectable with the available means.

(d) The observations cited under *a*, *b*, *c* indicate that air that is irradiated with x-rays has acquired the property of discharging electrified bodies with which it comes in contact.

(e) If this is really the case and in addition if the air retains this property for some time after being exposed to x-rays, it should be possible to discharge electrified bodies that themselves are not directly irradiated by x-rays simply by conducting irradiated air to them.

One can be convinced of the validity of this conclusion in different ways. I should like to describe one experimental set-up, although it is not the simplest one.

I used a brass tube 3 cm wide and 45 cm long; a few centimeters from one end of the tube, part of its wall was cut away and replaced with a thin sheet of aluminum; through the other end a brass sphere, fastened to a metal rod and insulated, was sealed air-tight into the tube. Between the sphere and the closed end of the tube there was soldered a little side tube, which could be connected to an exhaust apparatus; when suction was applied, air that passed the aluminum window on its way through the tube flowed around the brass sphere. The distance from window to sphere was over 20 cm.

I set this tube up inside the zinc box so that through the aluminum window of the tube the x-rays could enter perpendicularly to its axis and so that the insulated sphere lay in the shadow beyond the range of these rays. The tube and zinc box were connected to each other; the sphere was connected to a Hankel electroscope.

It was then observed that a charge either positive or negative given to the sphere was not influenced by the x-rays as long as the air remained at rest in the tube, but that at once the charge decreased considerably if irradiated air was drawn past the sphere by strong suction. When a constant potential from a storage battery was applied to the sphere and when irradiated air was continuously sucked through the tube, an electric current was produced just as if the sphere had been connected to the tube wall by a poor conductor.

(f) The question arises in what manner air can lose the property given to it by x-rays. Whether in time it loses the property itself, that is, without coming in contact with

other bodies, is still unsettled. However, it is certain that
a brief contact with a body that has a large surface and is
not necessarily electrified may render the air ineffective.
If, for example, one placed a sufficiently large stopper of
cotton so far into the tube that irradiated air must pass
through the cotton before it reaches the electrified sphere,
the charge of the sphere remains unchanged; even while
suction is applied.

If the stopper is placed in front of the aluminum
window, one obtains the same result as without cotton:
a proof that dust particles cannot possibly be the cause
of the discharge observed.

Wire screens have an action similar to cotton; however,
the screen must be very fine, and many layers must be put
on top of one another if the irradiated air passing through
them is to be made ineffective. If these screens are not
grounded, as has been assumed so far, but are connected
to a source of electricity of constant potential, the observations have always been what I anticipated; however, these
experiments have not yet been completed.

(g) If the electrified bodies are placed in dry hydrogen
instead of air, they are also discharged by x-rays. It seemed
to me that the discharge in hydrogen proceeded somewhat
slower; however, this is still uncertain because of the
difficulties in obtaining equal intensities of x-rays in a series
of consecutive experiments.

The method of filling the apparatus with hydrogen very
likely precludes the possibility that the denser layer of air
originally present on the surface of the bodies could play
an important role in the discharge.

(h) In highly evacuated spaces the discharge of a body
struck directly by x-rays proceeds much more slowly—in
one case, for example, about seventy times more slowly—

than in the same vessels when they are filled with air or hydrogen of atmospheric pressure.

(i) Experiments have been started on the behavior of a mixture of chlorine and hydrogen under the influence of x-rays.

(j) Finally, I should like to mention that one must often accept with caution the results of experiments on the discharging effects of x-rays in which the influence of the surrounding gas has not been taken into account.

19. In some cases it is advantageous to insert a Tesla apparatus (condenser and transformer) between the discharge apparatus, which furnishes x-rays, and the Ruhmkorff coil. This arrangement has the following advantages: First, the discharge tubes are less liable to be punctured and heat up less; secondly, the vacuum, at least so far as my home-made tubes are concerned, keeps for a longer time; and, thirdly, some apparatus produce more intense rays. Some tubes that were evacuated too little or too much to work satisfactorily on the Ruhmkorff coil alone functioned satisfactorily with the use of the Tesla transformer.

The question arises—and I should like, therefore, to mention it without contributing anything to its solution at present—whether x-rays can also be produced by a continuous discharge from a source of constant potential or whether fluctuations of the potential are absolutely necessary to produce them.

20. It is stated in paragraph 13 of my first communication that x-rays can be produced not only in glass but also in aluminum. In continuing the investigations along these lines no solid body could be found that was not able to produce x-rays under the influence of cathode rays. I also have found no reason for liquid and gaseous bodies' not acting in the same manner.

However, quantitative differences in the behavior of different bodies have been found. For example, if one lets cathode rays fall upon a plate, one half of which consists of a 0.3 mm platinum sheet and the other half of a 1 mm aluminum sheet, one observes on the photograph of this double plate taken with a pinhole camera that the platinum emits considerably more x-rays from the front side where it has been struck by the cathode rays than the aluminum emits from the same side. From the rear side, however, hardly any x-rays are emitted from the platinum but relatively many from the aluminum. In the latter, rays have been produced in the front layers of the aluminum and have penetrated through the plate.

One can easily arrive at an explanation of this observation, but it might be advisable to learn about some other properties of the x-rays first.

However, it should be mentioned that the observed facts also have a practical significance. According to my experience up to now, platinum is best suited for the production of x-rays of highest intensity. For several weeks I have used with good success a discharge tube with a concave mirror of aluminum as cathode and a platinum foil as anode, which has been placed in the focus of the cathode and inclined 45 degrees in relation to the axis of the mirror.

21. In this apparatus x-rays are emitted from the anode. From experiments made with apparatus of various shapes I must conclude that, insofar as the intensity of x-rays is concerned, it does not matter whether these rays are produced at the anode or not.

Especially for experiments with alternating currents from a Tesla transformer a discharge apparatus is being built, in which both electrodes are concave aluminum mirrors, whose axes form a right angle; in their common focus a platinum plate is placed that receives the cathode rays. A report on the usefulness of this apparatus will appear later.

Finished: March 9, 1896
Würzburg. Physikal. Institut d. Universität.

## Chapter VII
# March 1896 March 1897

HE TANGLE of reactions to his discovery, calling for his almost automatic responses to innumerable and varied demands upon his time and person, made it imperative that Röntgen get away from Würzburg. On the morning after he submitted his second communication, he and Mrs. Röntgen departed for a long anticipated holiday in Italy. Almost as fugitives they fled to Sorrento near Naples in search of uninterrupted rest. En route they found that his fame had traveled before him. In Munich he was recognized in the railroad station; he received and was obliged to decline an invitation to lecture on the rays. In Rome, where a newspaper reported that he had been seen wearing a brown suit, he was invited by the Italian physicists Salvioni and Righi to lecture at their universities. These invitations he also declined, and packing the brown suit in the bottom of his trunk, he immediately left for Sorrento. There at the beautiful Hotel Victoria, for the first time since the news of his discovery had enveloped him in a halo of personal greatness, he had leisure to gain perspective.

Two events occurring before they left Würzburg he now recalled with pleasure. On March 3 the University of Würzburg had made him an honorary doctor of medicine, this despite the pessimistic comments about the value of x-rays to medicine voiced by some of his colleagues, such as the surgeon Schönborn. Even more vivid in retrospect was the tribute from the Würzburg students, who had paraded with torches through the Pleicher Ring to the physical institute to acknowledge his achievement as a member of their university. To them he had said:

Fellow students: When I was young I had many ambitious aspirations; but my dreams have never gone so far as to imagine that the students of a great German university would ever have a torch-light parade in my honor to express their recognition for a purely scientific achievement. I express my heartfelt and sincerest thanks to you for this rare distinction and great honor, which I count among the highest given to me. I would like to add to my thanks a wish. As students of this university, you are especially chosen to take part in the future great progress of human knowledge, which is constantly advancing. To wish that every one of you might, at some time in his life, be honored by a parade in recognition of his scientific work would fit the mood of a day like this, but this hope is probably quite beyond the realm of possibility. If it should happen that one of you should be so honored, I would be very happy and would like to ask him to remember that I was the first to congratulate him. But instead of this rather far-reaching wish, I should like to give you another, the benefits of which I have tested myself.

During the time when congratulations and honors were showered upon me, unconsciously the new impressions erased the older ones, but one thought has always remained lively and fresh, and that is the memory of the satisfaction that I felt when my work was finally developed and completed. This is the joy derived from successful effort and from progress. You each can enjoy this happiness in life, and you each can and must reach this goal, which depends principally upon yourselves. May this happiness, this inner satisfaction, come to each of you, and may the circumstances permit you to attain this end by a path that is not too difficult.

This is the wish that I leave with you today. Now let me conclude by asking you to give three cheers for our beloved Alma Mater, the University of Würzburg.

In those days of leisure at Sorrento Röntgen's personal reaction to his discovery became more clearly crystallized, being perhaps best summarized in the words of Werner von Siemens, whom he had quoted the previous year upon assuming the rectorship of the university. In these words of Siemens' Röntgen found explained one's personal exultation in the discovery of a fundamental physical phenomenon such as the x-rays:

If some phenomenon that has been shrouded in obscurity suddenly emerges into the light of knowledge, if the key of a long sought mechanical combination has been found, if the missing link of a chain of thought is fortuitously supplied, this then gives the discoverer the exultant feeling that comes with a victory of the mind, which alone can compensate him for all the struggle and effort and lift him to a higher plane of existence.

Leaving Sorrento toward the end of March the Röntgens traveled north to the Bellevue Hotel in Cadenabbia at beautiful Lake Como. From Cadenabbia, which had become their favorite spring vacation spot, they made trips by steamer to Bellagio and to the Villa Arconati near Lenno, where from the old terraces there is a magnificent view of the lake. On the fifty-first anniversary of Röntgen's birth, after having secured a key to the Villa Carlotta from the Prince of Meiningen, they walked through the marvelous gardens of the villa, where the marble reliefs of Thorwaldson are superbly placed against the background of the plants and flowers. That evening at a dinner celebrating the day Röntgen received a sincere personal tribute from his old friend Krönlein, who as a young doctor had cared for him after a severe injury sustained in mountain-climbing. In his toast to Röntgen Krönlein enthusiastically supported von Kölliker's suggestion that the rays be named after their discoverer.

Although as a tribute from his friend the toast gave him a great deal of pleasure, Röntgen nevertheless had a strong personal objection to the identification of an individual's name with a natural phenomenon and on that occasion was amused by the thought that should Krönlein have been he, they might be talking about Krönlein rays or Krönlein-o-graphs. Throughout his life Röntgen was consistent in this stand and not only insisted upon calling the rays "x-rays" but also deeply resented the use of his name in association with them, particularly when this was done without his permission.

Before returning to Würzburg from a spring holiday in Italy, the Röntgens were accustomed to spending a few days in Baden-Baden. Accordingly upon leaving Cadenabbia they crossed from Italy by way of the St. Gotthard pass to Luzern, and thence to Zürich to visit friends and relatives. Then along the foot of the Black Forest, whose mountains in the golden light of the setting sun had an appeal for them equal to that of the Swiss mountains, they journeyed to Baden-Baden. There in an accumulation of mail, telegrams, and printed matter, forwarded from Würzburg to the Hotel de France, they were given a preview of the fast-moving public life they had escaped and to which they were returning.

In an impressively worded letter Mayor Sauerbronn of Lennep, the city of Röntgen's birth, informed him that the Lennep City Council, being proud that one of its sons had made a discovery of the greatest significance to German science, had passed a resolution making him an honorary citizen of the town. In reply Röntgen wrote on April 20, "I accept with warmest thanks and with great joy the high honor of becoming an honorary citizen of the city of Lennep."

Mrs. Röntgen at the time of the discovery of the roentgen rays.

Another indication of what might be expected in honors consequent to his gift to science appeared in a letter from a Munich friend, who stated confidentially that Röntgen would receive from the Prince Regent Luitpold of Bavaria the Royal Order of Merit of the Bavarian Crown, carrying with it nobility. The receiving of personal nobility had no meaning for Röntgen, and his stand on the matter was clearly stated some months later in a letter to the Mayor of Lennep. In replying to an announcement that representatives of the city of Lennep would personally present him with his diploma of honorary citizenship, Röntgen first expressed his anticipated pleasure in receiving them and invited them to be his guests, then he continued:

In regard to the prefix *von*, which you used in connection with my name in your letter, I should like to say the following: The paragraph of the rules of the Royal Bavarian Order of the Crown referring to this states: "The decoration of citizens with the Order of Merit of the Bavarian Crown includes the bestowal of personal nobility. The rights of nobility, including the use of the prefix of nobility *von*, can be exercised only after matriculation has been accomplished. Failure to make application for matriculation means forfeiting the rights of nobility." Since to date I have not made such an application, and since I am not intending to do so, I am not entitled to the use of the prefix *von*.

A reminder that administrative duties also awaited him as rector of the university in Würzburg was incorporated in a letter dated March 30 from Boveri: "However, we beg you urgently to be back not later than April 19 if it is at all possible. Your influence as rector . . . seems to us to demand absolutely your presence." This, with certain pangs of conscience, Röntgen laid aside; April 19 had already passed.

Perhaps the most legitimate measure of the importance of the discovery of x-rays and of the extent of its dissemination was to be found in the demand for reprints of the first communication. In three months five editions had been published. The first edition was the now classic one by "Dr. W. Röntgen" published "Ende 1895." The second edition appearing a few days later, early in 1896, listed the author as Dr. Wilhelm Konrad Röntgen, although Röntgen had always, and notably in his original manuscript, spelled the Conrad with a *C*. The third edition had a light brown wrapper and a separate title page with the same text as the wrapper. The fourth edition was identical with the third. In the fifth edition this statement appeared on the front cover: "The present booklet is also published in English, French, Italian, and Russian." Many of the later editions appeared with a wide, bright red strip announcing: "Contains the new discovery of Professor Röntgen of Würzburg."

Copies of his second communication, *On a New Kind of Rays, Continued,* had been forwarded to Baden-Baden. Stahel had also printed these and had bound them in an orange colored wrapper. This communication had a separate title page, carrying the same text as the cover and the price, 60 pfennig. Pages 3 to 9 contained the text, and three additional pages carried advertisements of the publisher. One of these ads announced: "X-ray picture of the hand of Geheimrat von Kölliker. Price 50 pfennig. This picture is of special interest since it was made by Professor Röntgen himself at that memorable meeting on January 23, 1896, in which he presented his discovery and also since this is the hand of the famous anatomist von Kölliker." Another stated: "In its 5th edition appeared Dr. Wilhelm Konrad Röntgen, Eine Neue Art von Strahlen I. Mitteilung, Preiss 60 pfennig."

Another of the seemingly endless ramifications of the discovery of the x-rays was the human interest and comic treatment, and Röntgen's mail was heavy with popular magazines containing cartoons and stories. Cartoons had appeared in the British *Punch* for January 25 and March 7 and in the American magazine *Life* for February 27 and March 26. Also in *Life* for March 12, 1896 was a fairly representative example of how the discovery was being celebrated in verse:

> She is so tall, so slender, and her bones—
> Those frail phosphates, those carbonates of lime—
> Are well produced by cathode rays sublime,
> By oscillations, amperes and by ohms.
> Her dorsal vertebrae are not concealed
> By epidermis, but are well revealed.
>
> Around her ribs, those beauteous twenty-four,
> Her flesh a halo makes, misty in line,
> Her noseless, eyeless face looks into mine,
> And I but whisper, "Sweetheart, je t'adore."
> Her white and gleaming teeth at me do laugh.
> Ah! Lovely, cruel, sweet cathodagraph!

And in *Photography:*

> The Roentgen Rays, the Roentgen Rays,
> What is this craze:
> The town's ablaze
> With the new phase
> Of x-ray's ways.
>
> I'm full of daze,
> Shock and amaze;
> For nowadays
> I hear they'll gaze
> Thro' cloak and gown—and even stays,
> These naughty, naughty Roentgen Rays.

Even the channels of trade and law had been affected: In London a firm advertised in February 1896 the "sale of x-ray-proof underclothing," and in the United States Assemblyman Reed of Somerset County, New Jersey, introduced a bill into the state legislature prohibiting the use of x-rays in opera glasses in the theaters. Many persons reacted strongly to the ghost pictures. The editor of the *Grazer Tageblatt* had a roentgen picture taken of his head and upon seeing the picture "absolutely refused to show it to anybody but a scientist. He had not closed an eye since he saw his own death's head."

Not a few letters in the mail forwarded to Baden-Baden from Würzburg intimated that there might be monetary gains in the proper exploitation of the rays. Some time later Max Levy, an engineer of the German electric firm A. E. G., who himself had done some excellent work with roentgen rays, approached Röntgen regarding his firm's interest in the development of the rays. Röntgen answered him without hesitation: "According to the good tradition of the German university professors, I am of the opinion that their discoveries and inventions belong to humanity and that they should not in any way be hampered by patents, licenses, contracts, nor should they be controlled by any one group."

An opposite yet tenable point of view was Edison's, who freely admitted to the commercial exploitation of science for personal gain. He was quoted by the American newspapers as saying:

Professor Röntgen probably does not draw one dollar profit from his discovery. He belongs to those pure scientists who study for pleasure and love to delve into the secrets of nature. After they have discovered something wonderful, someone else must come to look at it from the commercial point of view. This will also be the case with Röntgen's discovery. One must see how to use it and how to profit by it financially.

In a similar vein the editor of the *British Journal of Photography* wrote on February 21, "If the now famous roentgen ray skiagraphs had been made copyright, a very large sum could have been obtained for permission to use them as soon as the rays were 'boomed,' and what a boom there is and has been."

It was almost as though Röntgen had provided the newspaper world with an inexhaustible supply of hot copy. One sensational news story followed on the heels of the last. The *Electrical World* reported:

Edison himself has been having a severe attack of Röntgenmania. The newspapers having reporters in attendance at his laboratory do not suffer for copy, as the yards of sensational matter emanating from this source attest, and we learn that last week Mr. Edison and his staff worked through seventy hours without intermission, a hand organ being employed during the latter hours to assist in keeping the force awake.

New York newspapers stated:

At the College of Physicians and Surgeons the roentgen rays were used to reflect anatomic diagrams directly into the brains of the students, making a much more enduring impression than the ordinary methods of learning anatomical details.

Others thought that with the x-rays base metals could be changed into gold, vivisection outmoded, temperance promoted by showing drunkards the steady deterioration of their systems, and the human soul photographed.

Upon his return to Würzburg early in the summer of 1896, Röntgen was brought face to face with an entirely unsuspected development, claims to priority not only in investigation of uses of the rays but also in the discovery. The maiden issue of the English *Archives of Clinical Skiagraphy*, the first journal to be devoted to x-ray photography, edited by Sidney Rowland, B.A., Camb., "Special Commissioner to *British Medical Journal* for Investigation and Application of the New Photography to Medicine and Surgery," was published in May 1896 by the Rebman Publishing Company, Strand, in London and carried this comment by the editor:

Although Professor Röntgen's discovery is only a thing of yesterday, it has already taken its place among the approved and accepted aids to diagnosis. . . . The greater part of the practical improvements that have led to the present stage of perfection of the process have been made in this country. . . .

With regard to priority claims to the actual discovery, it is a fact that both Goodspeed of Philadelphia and Sir William Crookes had had freak experiences with photographic plates in their experiments with cathode rays. However, until Röntgen's discovery the phenomena remained unexplained. Goodspeed had actually made an accidental x-ray photograph on February 22, 1890. He had thrown it into a collection of freak photographic orphans, only to dig it out for evaluation over five years later when Röntgen announced his discovery. In a lecture on roentgen rays given at the University of Pennsylvania on February 22, 1896 Goodspeed concluded with the story of his early experiments:

We can claim no merit for the discovery, for no discovery was made. All we ask is that you remember, gentlemen, that six years ago, day for day, the first picture in the world by cathodic rays was taken in the physical laboratory of the University of Pennsylvania.

Crookes, according to reports, had returned some photographic plates to the manufacturer with the complaint that they were badly fogged. It had not occurred to him that experimenting with the Crookes tube near the plates might have had something to do with the plates' becoming darkened.

The speed with which the discovery was publicized and the immediate use of x-rays in photography might account for the haziness surrounding diversified claims of priority. Within a few days of the announcement of the discovery, Spies in Berlin, König in Frankfort o. M., Becher in Berlin, and Voller in Hamburg produced excellent x-ray pictures. Pioneers in x-ray photography in Vienna were the Exner brothers and Hascheck, and in February 1896 Eder and Valenta of Vienna published a portfolio mostly of animal x-ray pictures reproduced by a special technic giving them the appearance of delicate engravings. Also from Vienna a picture of a hand was sent to Röntgen made by Hascheck at the request of his colleague Lindenthal. The veins had been injected with Teichmann's mixture of lime, cinnabar, and petroleum via the brachial artery and stood out in beautiful relief. This effect produced by injecting blood vessels suggested the possibility of other organs of the human body, such as the esophagus, stomach, intestines, lungs, gallbladder, and brain being filled with heavier substances so that by contrast on the x-ray plate the organ and its function would be revealed.

In England early x-ray pictures were made by Schuster, Stanton, Thomson, Lodge, Swinton, Edwards, and Rowland. Among the first to use x-rays in photography in France were Barthélemy, Oudin, Benoist, Guillaume, Séguy in Le Roux's laboratory, and Lannelongue and Londe in the Salpêtrière Hospital. Among those who sent

Röntgen x-ray photographs from the United States were Miller of Case School, Frost of Dartmouth College, Goodspeed, Wright, and Trowbridge of Harvard University, Edison, and Norton. From Pupin of Columbia University he received a picture of a hand full of shot, in which a remarkable detail had been obtained by placing a fluorescent screen behind the plate while making the x-ray, the precursor of the intensifying screen.

During the summer of 1896 Röntgen was visited by scientists from all over the world, who came to compare notes and to exchange with him experiences with the rays. Among these was Dayton C. Miller from Cleveland, who had bought the entire exhibit of Crookes and Geissler tubes displayed by the Geissler firm at the Chicago Exposition of 1893 and thus was well equipped to experiment with the rays when the discovery was announced. He brought Röntgen greetings from his cousin Johann Heinrich Röntgen, minister of the First Reformed Church in Cleveland and superintendent of the Bethesda German Hospital, and his cousin Louise. He also spoke of early collaboration with the surgeon George Crile and the dentist Weston A. Price and gave Röntgen a picture of an entire human skeleton fitted together in sections and made as early as March 1896. Miller acquired several of the newly developed tubes with platinum targets to take back with him.

Likewise during the summer of 1896 Röntgen was flooded with invitations to lecture on his discovery. He repeatedly refused the invitation of the British Association for the Advancement of Sciences to speak before their session in September and instead made plans for another holiday. Early in August he and Mrs. Röntgen headed for Switzerland with the intention of attending sessions of the Swiss Congress of Scientists in Zürich. The same problem that had irritated them on their southern holiday in the spring, of being the object of public curiosity, had to be faced in Zürich, and finally to escape the crowds they moved on to meet the Zehnders in Rigi-Scheidegg. Again Röntgen found himself surrounded by crowds and faced by cameras, and again they beat a retreat, this time to the Weisses Kreuz in Pontresia to find protection from intrusions under the roof of their old friends the Enderlins. That a great change had come over Röntgen was very evident, particularly to young Trippi, Enderlin's grandson. On former visits to Pontresia Röntgen had carried a camera with him everywhere. Now he was almost camera-shy.

When bad weather speeded the Röntgens' return to Würzburg after a fortnight in the Engadin and a few days in Weissbad in the Swiss Appenzell, Röntgen was ready to go to work in earnest and began a series of experiments on the increase in the penetration of x-rays with increasing voltages at the tube. This study was attended by certain obstacles. As soon as the voltage applied reached a certain point, the tubes were punctured, and Röntgen's budget at the university was by no means high enough to include the purchase of the numerous tubes needed in these experiments. Accordingly when Rosenthal, engineer of Reiniger-Gebbert and Schall, sent him tubes that behaved splendidly even under his exacting requirements, Röntgen, despite great professional pride making it difficult for him to ask favors, wrote in a letter dated November 27, 1896:

> Your tubes are really very good, but too expensive for my means. I use these tubes not only for the well known experiments but also as is apparent for many other experiments in

which the tubes must stand a much greater strain than normally, and the result is that they are destroyed more quickly. I would like to ask you whether you are able to let me have the tubes for twenty instead of thirty marks. . . .

Almost as rapid as the use of x-rays in photography was the development of the x-ray equipment industry. But even in his most fantastic dreams Röntgen could not possibly have visualized the construction fifty years later of tubes nine feet long operating at several million volts and producing x-rays able to penetrate a foot of steel, such a tube costing thousands of dollars. Conical Hittorf tubes of the sort Röntgen used at the time of his discovery were advertised in May 1896 by expert glassblowers in Gehlberg, Thuringia, for about two dollars, and improved "special" tubes with a platinum anode for less than four dollars. Ferdinand Ernecke of Berlin offered large Hittorf tubes for six dollars; a four by eight inch barium platinocyanide screen for two dollars; a storage battery with current regulator, an instruction book on photography, and a large Ruhmkorff induction coil with a six inch spark gap for less than one hundred and fifty dollars. A pretentious ad by Siemens and Halske, published in the *Zeitschrift für Electrochemie* in March 1896, offered induction coils for six inch gap, mercury interrupter, and two glass spheres for two hundred dollars. In June 1896 the F. J. Pearson Manufacturing Co. of St. Louis advertised a "portable x-ray apparatus for physicians, professors, photographers, and students, complete in handsome case, including coil, condenser, two sets of tubes, battery, etc., for the price of $15 net delivered in the United States with full guarantee."

Many other firms advertised roentgen equipment, some of which are now out of existence: Geissler in Berlin, Müller in Hamburg, Poeller in Munich, Goetze in Leipzig,

Greiner & Friedrichs in Stützerbach; in France Heller and Radiguet in Paris; in England Thomas and Griffin in London; in Italy Gorla in Milano; in the United States Edison Decorative and Miniature Lamp Department in Harrison, N.J., Beacon Lamp Co. and Knott in Boston, Sunbeam Lamp Co. in Chicago, Willyoung in Philadelphia, and Greiner in New York.

That reasonably priced x-ray equipment was so quickly available was in no small part due to Röntgen's refusal to restrict the development of his discovery in any way, and there is no evidence that the discovery of the rays was exploited commercially. Very few patents on x-ray apparatus were applied for in 1896. One German electrical firm, one engineer, and one scientist applied for protection of their ideas on x-ray tubes, and one French scientist and six English and American x-ray workers applied for patents on other phases of the discovery. A few publishers also tried to obtain copyright protection for early x-ray pictures.

One aspect of the wide use of the x-rays perturbed Röntgen greatly. From laboratories in the United States, England, Germany, and France came more and more reports of a peculiar skin reaction similar to a sunburn in persons working with the rays, some of these reactions being particularly serious. In New York Mr. Hawks, a demonstrator of roentgenography in Bloomingdale Brothers department store, suffered a very severe reaction. His apparatus, which was running continuously for two or three hours every day, was described in the *Electrical Engineer* in July 1896:

> To add to the many attractions of their large establishment, Bloomingdale Brothers have recently opened an x-ray exhibition. A few words as to the details of this apparatus may be of interest. The coil is an eight-inch Splitdorf with the make and

break circuit mounted on the shaft of a motor, the spark at the break being absorbed by condensers. The tube is of the focusing type and is made by Greiner of this city. By proper manipulation I am able to make a very clear photograph of the hand with 20 to 30 seconds' exposure and a picture of the ribs in about 10 to 15 minutes. A very essential thing in running tubes to the maximum effect is to keep the air around them dry. All who are interested in the x-ray should call at Bloomingdale Brothers and see the apparatus in operation.

The first effect Hawks observed was a slight dryness of the skin, which increased and soon resembled a very strong sunburn. The nails of his hands stopped growing, and skin areas exposed to the x-rays began to scale. In demonstrating the rays' penetration of the skull he had placed his head close to the tube, with the result that his vision was impaired and that the hair of the temples, eyebrows, and eyelashes came out. Sunburn-like effects were also observed on the chest. The pain attendant to these effects was so severe that he required medical attention, and to protect himself he covered his hands with vaseline and used gloves. This was to no avail.

It distressed Röntgen to believe that these effects were due to x-rays. Other scientists suggested that they might be due to the effects of ultraviolet radiation, to the impact of minute platinum atoms from the target, to the effects of cathode rays, to electrical induction currents, to ozone generated in the skin, to an idiosyncracy of the irradiated individual, or possibly to faulty technic. The most acceptable explanation to Röntgen was that of Sir Joseph Lister, who opened the Liverpool Congress of the British Association for the Advancement of Science in September 1896 and said in his presidential address:

There is another way in which the roentgen rays connect

themselves with physiology and may possibly influence medicine. It is found that if the skin is long exposed to their action it becomes very much irritated, affected with a sort of aggravated sunburning. This suggests the idea that the transmission of the rays through the human body may be not altogether a matter of indifference to internal organs, but may by long continued action produce, according to the condition of the part concerned, injurious irritation or salutary stimulation.

Röntgen almost from the beginning had conducted all his experiments in his big zinc box better to define his x-ray beam by means of diaphragms in the box and also to protect the photographic plates. Then he had added a lead plate to the zinc between the tube and himself, and in doing so he had unknowingly protected himself completely. With an accurate appreciation of the import of these unexplained effects, Röntgen foresaw the suffering consequent to the careless handling of the rays.

Also of great interest were the first x-ray evidences before court. One case, which was described in detail on March 20, 1896 in the *British Journal of Photography* and in June in the *British Medical Journal*, was commented upon in American journals as early as April 1896 (*Literary Digest*, April 11, 1896, and *Electrical Engineer*, New York, June 10, 1896). The *Literary Digest* wrote, for instance, under the title *The New Photography in Court:*

An interesting and novel case, in which the "X-rays" practically decided the point, was tried by Mr. Justice Hawkins and a special jury at Nottingham the other day, says *The Hospital*, London. Miss Ffolliott, a burlesque and comedy actress, while carrying out an engagement at a Nottingham theater early in September last, was the subject of an accident. After the first act, having to go and change her dress, she fell on the staircase leading to the dressing-room and injured her foot. Miss Ffolliott remained in bed for nearly a month, and at the end of that time was still unable to resume her vocation. Then, by the advice of Dr. Frankish, she was sent to University College Hospital, where both her feet were photographed by the "X-rays." The negatives taken were shown in court, and the difference between the two was convincingly demonstrated to the judge and jury. There was a definite displacement of the cuboid bone of the left foot, which showed at once both the nature and the measure of the injury. No further argument on the point was needed on either side, and the only defense, therefore, was a charge of contributory carelessness against Miss Ffolliott. Those medical men who are accustomed to dealing with "accident claims"—and such claims are now very numerous—will perceive how great a service the new photography may render to truth and right in difficult and doubtful cases. If the whole osseous system, including the spine, can be portrayed distinctly on the negative, much shameful perjury on the part of a certain class of claimants, and many discreditable contradictions among medical experts, will be avoided. The case is a distinct triumph for science, and shows how plain fact is now furnished with a novel and successful means of vindicating itself with unerring certainty against opponents of every class.

The trend of the reaction to the discovery in medicine, pure science, industry, the courts, and the press could in no wise be changed, although the good far overbalanced the disagreeable. Yet more basic work, colorless as it may have seemed by comparison, had to be done, and, in spite of unavoidable dissipation of Röntgen's energies, his findings for the third communication were beginning to take shape at the beginning of the year 1897. Any further publication of observations on properties of x-rays had been requested by the Prussian Academy of Sciences for their communications. Although he would have preferred to have his third communication follow the first two in

the *Sitzungsberichte* of the Würzburg Society, he realized
that he was not free to make that decision, and on March
10, 1897 he sent his third communication on x-rays to the
editor of the reports of meetings of the Prussian Academy
of Sciences in Berlin. This communication was the first
one of the three to have an illustration, a line drawing of
a simple apparatus. In none of his communications did
Röntgen present an x-ray picture.

## Chapter VIII*

# March 10, 1897

### W. C. Röntgen: Further Observations on the Properties of X-rays

### (Third Communication)

1. If one places an opaque plate between a discharge apparatus[1] that emits intense x-rays and a fluorescent screen in such a way that the shadow of the plate covers the entire screen, one can still detect a luminosity of the barium platinocyanide. This light can even be seen if the screen lies directly upon the plate, and one is at first inclined to think that the plate is transparent. However, if one covers the screen lying on the plate with a heavy plate of glass, the fluorescent light becomes much weaker and disappears entirely if instead of a glass plate one places the screen in a cylinder of leadfoil 0.1 cm thick, which is closed at one end with the opaque plate and at the other by the head of the observer.

The phenomenon described may have been produced by diffraction of rays of very long wavelength or by the fact that x-rays are emitted from substances surrounding the discharge apparatus, notably from the irradiated air.

* This new translation of Röntgen's third communication varies somewhat from one made by G. F. Barker in 1897.

[1] All the discharge tubes mentioned in the following communication are constructed according to the principle given in paragraph 20 of my second communication (Sitzunsber. d. phys.-mediz. Gesellschaft zu Würzburg, Jahrg. 1895). I obtained a great number of them from the firm of Greiner and Friedrichs in Stützerbach i. Th., to whom I wish to express publicly my thanks for putting abundant material at my disposal gratis.

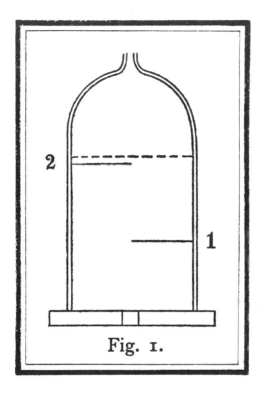

Fig. 1.

The latter explanation is the correct one as can be easily demonstrated with the following apparatus, among others. Figure 1 represents a very thick-walled glass bell jar, 20 cm high and 10 cm wide, which is closed and sealed with a heavy zinc plate. At 1 and 2 are inserted circular segments of lead sheets, which are somewhat larger than half the cross section of the jar and which prevent the x-rays that enter the jar through an opening in the zinc plate, which is covered with a celluloid film, from travelling in a straight line to the space above lead disk 2. On the upper side of this lead disk is fastened a small barium platinocyanide screen, which almost fills the entire cross section of the jar. This cannot be hit by direct rays nor by those which have undergone a primary diffuse reflection on a solid substance (for example, the glass wall). The jar is filled with dust-free air before each experiment.—If one lets x-rays enter the jar, first, so that they are all stopped by lead screen 1, one does not yet see any fluorescence near 2; only when the jar is tipped so that direct rays can also enter the space between 1 and 2 does the fluorescent screen show an illumination of the half not covered with lead disk 2. If the jar is then connected to a water aspirator, one notices that the fluorescence becomes gradually weaker as the evacuation progresses; if air is readmitted, the intensity increases again.

Since now, as I found, mere contact with air that has just been irradiated does not produce any noticeable fluorescence of the barium platinocyanide, one must conclude from the experiment described that air emits x-rays in all directions while it is being irradiated.

If our eye were as sensitive to x-rays as it is to light rays, a discharge apparatus in operation would appear to us like a light burning in a room that is uniformly filled with tobacco smoke; perhaps the color of the direct irradiation and that coming from the air particles would be different.

I have not yet been able to answer the question as to whether the rays that are emitted from irradiated substances are of the same kind as those impinging upon them or, in other words, whether a diffuse reflection or a phenomenon similar to fluorescence is the cause of these rays; that the rays coming from the air also are effective photographically can easily be proved; as a matter of fact, this effect is even noticeable sometimes in a manner disagreeable to the observer. To guard against this, which is frequently necessary, especially for longer exposure times, one must enclose the photographic plate in suitable lead containers.

2. For comparing the intensity of the radiation of two discharge tubes and for several other experiments I used an arrangement that is fashioned after the Bouguer photometer, and that I shall also simply call a photometer. A rectangular sheet of lead, 35 cm high, 150 cm long, and 0.15 cm thick, is placed vertically at the center of a long table and supported by boards. On each side of it is placed a discharge tube which can be moved along the table. At one end of the lead strip a fluorescent screen[2] is attached in such a way that each half of it receives perpendicularly the rays from only one of the tubes. In these measurements one adjusts to obtain equal intensity of the fluorescence in both halves.

Some remarks on the use of this instrument may be made here. First, it must be stated that adjustments are frequently very difficult to make because of the incon-

[2] In this and in other experiments Edison's fluorescent screen has proved very useful. This consists of a box similar to a stereoscope, which can be held light-tight against the head of the observer and whose cardboard end is covered with barium platinocyanide. Edison uses tungstate of calcium instead of barium platinocyanide, but I prefer the latter for several reasons.

stancy of the source of radiation; the tube responds to each irregularity in the interruption of the primary current, such as occurs with the Deprez and notably with the Foucault interrupter. It is therefore advisable to make repeated adjustments.

Secondly, I should like to indicate the factors that govern the brightness of a given fluorescent screen which is bombarded by x-rays in such rapid succession that the observing eye can no longer detect the intermittence of the radiation. This brightness depends upon (1) the intensity of the radiation emitted from the platinum plate of the discharge tube; (2) very probably the kind of rays that fall upon the screen, since not every type of radiation causes the same degree of fluorescence (see below); (3) the distance of the screen from the point of emission of the rays; (4) the absorption of the rays on their way to the barium platinocyanide; (5) the number of discharges per second; (6) the duration of each single discharge; (7) the duration and the strength of the afterglow of the barium platinocyanide; and (8) the radiation originating in materials surrounding the discharge tube and falling upon the screen. In order to avoid mistakes one should always remember that in general these conditions are similar to a comparison of the fluorescent action produced by two intermittent light sources of different colors which are surrounded by an absorbing envelope and placed within a turbid—or fluorescent—medium.

3. According to paragraph 12 of my first communication[3] the part of the discharge apparatus that is struck by cathode rays is the point of emission of x-rays which spread out "in all directions." Now it is interesting to learn how the

[3] Sitzungsberichte der physik.-mediz. Gesellschaft zu Würzburg. Jahrg. 1895.

intensity of the rays varies with the direction. For this investigation the sphere-shaped discharge apparatus with smoothly polished plain platinum plates upon which the cathode rays fall at an angle of 45 degrees are the most suitable. Even without additional instruments one can recognize from the uniformly bright fluorescence of the hemispherical glass wall above the platinum plate that there are no very great variations in the intensities in different directions and that therefore *Lambert's* law of emission cannot hold here; nevertheless, this fluorescence might still be produced largely by cathode rays.

In a more accurate test the intensity of the radiation emitted in different directions from several tubes was examined with the photometer; furthermore for the same purpose I have exposed photographic films bent in the shape of a semicircle (radius 25 cm) with the platinum plate of the discharge apparatus as its center. In both procedures, however, the variation in thickness of different areas of the tube wall becomes very disturbing, since it causes x-rays proceeding in different directions to be absorbed to different degrees. However, it seems entirely feasible to equalize the thickness of the glass through which the rays pass by interposing thin glass plates.

The result of these experiments is that the radiation through an imaginary hemisphere with the platinum plate as its center is practically uniform almost to its very edge. Only when the angle of emission of the x-rays reached about 80 degrees could I detect the beginning of a decrease in the radiation, but even this decrease is still relatively small, so that the main variation in the intensity occurs between 89 and 90 degrees.

I have not been able to observe a difference in the kind of rays emitted a different angles.

On account of the described distribution of intensity of the x-rays, images from the platinum plate observed either upon the fluorescent screen or upon the photographic plate by means of a pinhole camera—or with a narrow slit—must be more intense the greater the angle between platinum plate and screen or photographic plate, provided that this angle does not exceed 80 degrees. I was able to confirm this conclusion by means of suitable arrangements which permitted comparisons of images obtained simultaneously at different angles from the same discharge tube.

In optics we encounter in the case of fluorescence a similar distribution of intensity of emitted radiations. If one adds a few drops of fluorescein solution to water in a rectangular tank and if one illuminates the tank with white or violet light, one observes that the brightest fluorescence proceeds from the edges of the slowly dropping threads of fluorescein, that is, from those parts where the angle of emission of the fluorescent light is greatest. Mr. Stokes on the occasion of a similar experiment has already explained that this phenomenon is due to the fact that rays which excite fluorescence are absorbed to a much greater extent by the fluorescein solution than is the fluorescent light itself. Now it is most remarkable that also cathode rays, which produce x-rays, are absorbed by platinum to a much greater extent than are x-rays, and the postulate suggests itself that a relationship exists between the two phenomena—the transformation of light into fluorescent light and of cathode rays into x-rays. However, at present no definite evidence for such an assumption exists.

Also the observations on the intensity distribution of the rays emitted from the platinum plate have a certain significance with respect to the technic of producing shadow pictures with x-rays. According to the statements made previously it is advisable to place the discharge tube in such a position that the rays used to produce the picture leave the platinum at the greatest possible angle, although it should not be much greater than 80 degrees; in this way one obtains the sharpest picture possible, and if the platinum plate is very plain and if the tube has been constructed so that the oblique rays do not have to pass through a glass wall considerably thicker than do the rays that are emitted perpendicularly to the platinum plate, then the radiation falling upon the object in the described arrangement does not suffer a decrease in intensity.

4. In my first communication I designated "transparency of a material" as the ratio of the brightness of a fluorescent screen placed perpendicular to the rays directly behind the material to that of the screen under identical conditions without interposition of the material. Specific transparency of a substance will be used to indicate the transparency relative to the unit thickness of the substance; this is equal to the $d$th root of the transparency when $d$ is the thickness of the traversed layer measured in the direction of the rays.

Since my first communication I have used mainly the photometer described previously to determine the transparency. A plate of the substance to be investigated— aluminum, tin, glass, and so forth—was placed in front of one of two equally bright fluorescent halves of the screen, and the difference in the brightness thus produced was then matched, either by increasing the distance between the discharge apparatus and the uncovered half of the screen or by bringing the other one closer. In both cases the correctly determined ratio of the squares of the distances of the platinum plates of the discharge apparatus from the screen before and after adjustment of the apparatus represents the desired value for the transparency of the interposed sub-

stance. Both methods led to the same result. After adding
a second plate to the first, one finds in the same way the
transparency of that second plate to the rays that have already passed through the first.

The described procedure presupposes that the brightness
of a fluorescent screen is inversely proportional to the square
of the distance from the source of radiation, and this is only
true if, first, the air does not absorb or emit any x-rays and,
secondly, if the brightness of the fluorescent light is proportional to the intensity of the radiation for rays of the
same kind. Now, the first condition certainly is not fulfilled,
and it is doubtful whether the second one is; I therefore
first convinced myself by experiments, as already described
in paragraph 10 of my first communication, that deviations
from the law of proportionality mentioned before are so
small that they may be neglected in our case.—Considering
the fact that x-rays are also emitted from irradiated substances, it should also be mentioned that, first, no difference
could be detected with the photometer in the transparency
of an aluminum plate, 0.925 mm thick, and of thirty-one
aluminum foils, each 0.0299 mm thick stacked on one another—31 times 0.0299 equals 0.927; and, secondly, that the
brightness of the fluorescent screen was not noticeably
different when the plate was placed directly in front of the
screen or at a greater distance from it.

The result of these transparency experiments for aluminum is as follows:

| Transparency to perpendicularly impinging rays | Tube 2 | Tube 3 | Tube 4 | Tube 2 |
|---|---|---|---|---|
| Of the first 1 mm thick Al plate | 0.40 | 0.45 | | 0.68 |
| Of the second 1 mm thick Al plate | 0.55 | 0.68 | | 0.73 |
| Of the first 2 mm thick Al plate | | 0.30 | 0.39 | 0.50 |
| Of the second 2 mm thick Al plate | | 0.39 | 0.54 | 0.63 |

From these experiments and from similar ones with glass and tin we arrive first at the following conclusion: If one assumes that the investigated substances are divided into layers of equal thickness, placed perpendicularly to the parallel rays, one sees that each of these layers is more transparent to the transmitted rays than the previous one, in other words: The specific transparency of a substance increases with its thickness.

This result is in complete agreement with what one observes on the photograph of a tinfoil ladder, as mentioned in paragraph 4 of my first communication, and also with the fact that occasionally on photographs the shadows of thin layers, such as, for example, of the paper used to wrap the plates, are sometimes very noticeable.

5. If two plates of different substances are equally transparent, this equality may not persist if the thickness of the two plates—but nothing else—is changed in the same ratio. This fact may be proved most simply with two scales, one of platinum and one of aluminum, placed side by side. For this purpose I used platinum foil, 0.0026 mm thick, and aluminum foil, 0.0299 mm thick. When I brought this double scale in front of the fluorescent screen or of a photographic plate and directed rays upon it, I found in one case, for example, that a single platinum layer was as transparent as a sixfold aluminum layer; however, the transparency of a twofold platinum layer was not equal to that of a twelvefold but to a sixteenfold aluminum layer. With another discharge tube I found that 1 platinum equals 8 aluminum and 8 platinum equals 90 aluminum. These experiments prove that the ratio of the thicknesses of platinum and aluminum of equal transparency is the smaller the thicker the respective layers are.

6. The ratio of the thicknesses of two equally transparent plates of different materials depends upon the thickness and the material of that substance—for instance, the glass wall of the discharge apparatus—that the rays must penetrate before they reach the respective plates.

In order to prove this conclusion—which is not un-expected according to statements made in paragraphs 4 and 5—one may use an arrangement that I call a platinum-aluminum window, which, as we shall see, is also useful for other purposes. It consists of a rectangular piece of platinum foil (4.0 cm by 6.5 cm), 0.0026 mm thick, which is glued to a thin paper screen and in which are punched 15 round holes, 0.7 cm in diameter, arranged in three rows. These little windows are covered with tightly fitting little disks of aluminum foil, 0.0299 mm thick, carefully stacked in such a way that there is one little disk in the first window, two in the second, and so forth, and finally fifteen disks in the fifteenth. If one places this arrangement in front of the fluorescent screen, one observes very clearly, particularly if one uses tubes that are not too hard (see below), the number of aluminum disks having a transparency equal to that of the platinum foil. This number will be called briefly the window-number.

In one case, when using direct radiation, I obtained the window-number 5; when a plate 2 mm thick made of ordinary soda glass was then interposed, the window-number obtained was 10; thus the ratio of the thicknesses of platinum and aluminum foil of equal transparency was reduced to one-half when I used rays that had passed through a glass plate 2 mm thick instead of rays that came directly from the discharge apparatus, q.e.d.

The following experiment should also be mentioned

here. The platinum-aluminum window was laid on a small package containing twelve photographic films and was then exposed; after development the first film lying under the window showed the window-number 10, the twelfth the number 13, and the others in proper sequence all the steps from 10 to 13.

7. The experiments described in paragraphs 4, 5, and 6 refer to the changes that the x-rays emitted from a discharge tube undergo in passing through different substances. It will now be proved that for one and the same substance and the same thickness traversed the transparency may be different for rays emitted from different tubes.

For this purpose the values for the transparency of an aluminum plate 2 mm thick for rays produced in different tubes are given in the following table. Some of these values have been taken from the first table in section 4.

| Transparency | Tube 1 | Tube 2 | Tube 3 | Tube 4 | Tube 2 | Tube 5 |
|---|---|---|---|---|---|---|
| For rays falling perpendicularly upon an aluminum plate 2 mm thick | 0.0044 | 0.22 | 0.30 | 0.39 | 0.50 | 0.59 |

The discharge tubes differ only slightly in construction or in the thickness of the glass wall but vary chiefly in the degree of evacuation of the gas content and in the discharge potential consequent to this; tube 1 requires the lowest, tube 5 the highest, discharge potential, or, as we shall say for the sake of brevity: Tube 1 is the softest, and tube 5 is the hardest. The same Ruhmkorff coil—directly connected to the tubes—, the same interrupter, and the same primary current were used in all cases.

All the many other materials that I investigated behave similarly to aluminum: All of them are more transparent to rays of a harder tube than to rays of a softer tube.[4] This fact seems to me worthy of special attention.

The ratio of the thicknesses of two equally transparent plates of different materials was also found to be dependent upon the hardness of the discharge tube used. One can recognize this immediately with the platinum-aluminum window (paragraph 5); using a very soft tube one obtains, for example, the window-number 2, while for very hard but otherwise identical tubes a scale, reading up to number 15, is not even sufficient. This means that the ratio of the thicknesses of platinum and aluminum of equal transparency is the smaller the harder the tubes are which emit the rays or—considering the result mentioned above—the less absorbable the rays are.

The different behavior of rays produced in tubes of different degrees of hardness is also evident, of course, in the familiar shadow pictures of hands, and so forth. Using a very soft tube one obtains dark pictures in which the bones are not very prominent; when a harder tube is used, the bones become clearly visible in all details, while the soft parts are weak in comparison, and with a very hard tube one obtains only weak shadows, even of the bones. From this observation one learns that the choice of the tube to be used must depend upon the nature of the object to be pictured.

8. It must also be mentioned that the quality of radiation emitted from one and the same tube depends upon different circumstances. As the investigation with the platinum-aluminum window shows, this is influenced: (1) by the manner in which the Deprez or Foucault interrupter[5] func-

---

[4] On the behavior of "non-normal" tubes see under 8.

[5] A good Deprez interrupter functions more uniformly than a Foucault interrupter; the latter, however, makes better use of the primary current.

tions in connection with the induction apparatus, that is, by the course of the primary current. Here must be mentioned the frequently observed phenomenon that some of the discharges in rapid succession produce x-rays that are not only particularly intense but that also are distinguished from others by their absorbability. (2) By a spark gap connected in series in the secondary circuit of the discharge apparatus. (3) By inserting a Tesla transformer in the circuit. (4) By the degree of evacuation of the discharge apparatus (as was mentioned previously). (5) By various, not yet sufficiently understood, phenomena in the interior of the discharge tube. Several of these factors deserve to be discussed in a little more detail.

If we take a tube that has not yet been used nor even evacuated and connect it to the mercury pump, we shall reach after the necessary pumping and heating of the tube a degree of evacuation in which the first x-rays may be noticed by a feeble light on a nearby fluorescent screen. A spark gap connected in parallel with the tube registers sparks a few millimeters in length, the platinum-aluminum window shows very low numbers, and the rays are very absorbable. The tube is "very soft." Now if a spark gap in series or a Tesla transformer is inserted,[6] more intense and less absorbable rays are produced. I found for instance in one case that by increasing the spark gap in series the window-number could gradually be brought from 2.5 up to 10.

(These observations prompted me to wonder whether x-rays might not be obtainable even at still higher pressures by using a Tesla transformer. This is indeed the case: Using

---

[6] That a spark gap connected in series acts similarly to a Tesla transformer I was able to point out in the French edition of my second communication (*Archives des Sciences Physiques*, etc., de Genève, 1896); in the German publication this comment was omitted by an oversight.

a narrow tube with wire-shaped electrodes I could still obtain x-rays when the pressure of the enclosed air amounted to 3.1 mm of mercury. If hydrogen was used instead, the pressure could even be higher. I was not able to determine the lowest pressure at which x-rays can still be produced in air; at any rate, it lies below 0.0002 mm of mercury, so that the range of pressures within which x-rays may altogether be produced is already now a very large one.)

As a result of further evacuation of the "very soft" tube—connected directly to the induction coil—the radiation becomes more intense, and a larger percentage of it passes through the irradiated material: A hand held in front of the fluorescent screen is more transparent than before, and higher window-numbers are obtained with the platinum-aluminum window. At the same time the spark gap connected in parallel must be increased in order to let the discharges pass through the tube: The tube has become "harder." If one evacuates the tube still more, it becomes so "hard" that the spark gap must be increased to beyond 20 cm, and now the tube emits rays to which the materials are exceedingly transparent: Heavy iron plates 4 cm thick, when investigated with the fluorescent screen, were still found to be transparent.

The behavior of a tube on the mercury pump connected directly to the induction coil, as described above, is normal, but deviations from this norm, which are caused by the discharges proper, occur frequently. Sometimes the behavior of the tubes is altogether unpredictable.

We have thought that the hardening of a tube is produced by continued evacuation with the pump, but it may also occur in a different way. A medium hard tube that has been sealed off the pump will gradually become harder by itself—unfortunately, as regards the duration of its usefulness—even when it is used correctly for producing x-rays, that is, when discharges are passed through it that do not or only faintly cause the platinum to glow. A gradual self-evacuation takes place.

With such a tube that had become very hard I obtained a very beautiful photographic shadow picture of the double barrel of a hunting gun with cartridges in place, in which all details of the cartridges, the internal faults of the damask barrels, and so forth, could be recognized very distinctly and sharply. The distance between the platinum plate of the discharge tube and the photographic plate was 15 cm, the time of exposure twelve minutes—which is comparatively long because of the smaller photographic effect of the less absorbable rays (see below). The Deprez interrupter had to be replaced by the Foucault interrupter. It would be interesting to construct tubes permitting the use of still higher discharge potentials than has been possible thus far.

The reason for the hardening of a tube that had been sealed off the pump was given above as self-evacuation caused by discharges; however, this is not the only cause, since changes taking place on the electrodes also have the same effect. What they consist of, I do not know.

A tube that has become too hard can be made softer by admitting air, sometimes also by heating the tube or reversing the direction of the current, and finally by sending very strong discharges through it. In the last case, however, the tube has for the most part acquired other properties than those described above: It sometimes requires, for example, a very high discharge potential and yet emits rays of a relatively low window-number and great absorbability. I do not wish to discuss further the behavior of these "non-normal" tubes.—The tubes constructed by Mr. Zehnder with an adjustable vacuum, since they contain a small piece of charcoal, have been very serviceable to me.

The observations described in this paragraph and others have led me to the conclusion that the composition of the rays emitted from a discharge tube equipped with a platinum anode depends primarily upon the duration of the discharge current. The degree of evacuation, the hardness, plays a role only because the form of discharge current depends upon it. If one is able to produce that form of discharge which is necessary for the production of x-rays by any form whatever, x-rays can also be produced even for relatively high pressures.

Finally it is worth mentioning that the quality of the rays produced by a tube is either not at all or only slightly changed when the strength of the primary current is altered considerably, provided that the interrupter functions the same in all cases. On the contrary the intensity of the x-rays is found to be proportional within certain limits to the strength of the primary current, as is demonstrated by the following experiment. The distances from the discharge apparatus at which in a certain case the fluorescence of the barium platinocyanide screen was barely noticeable amounted to 18.1 m, 25.7 m, and 37.5 m when the strength of the primary current was increased from 8 to 16 to 32 amp. The squares of those distances are in nearly the same ratio as the corresponding currents.

9. The results described in the last five paragraphs were obtained directly from the respective experiments mentioned above. If one surveys the sum of these individual results, one arrives, partly guided by the analogy that exists between the behavior of optical rays and x-rays, at the following impressions:

(a) The radiation emitted from a discharge apparatus

consists of a mixture of rays of different absorbability and of different intensity.

(b) The composition of this mixture depends essentially upon the time relationship of the discharge current.

(c) The rays which are selectively absorbed by various substances differ for different materials.

(d) Since x-rays are produced by means of cathode rays and since both have common properties—such as production of fluorescence, photographic and electrical effects, and absorbability, the amount of which depends essentially upon the density of the irradiated material, and so forth—, the hypothesis is suggested that both phenomena are processes of the same nature. Without being willing to adhere unconditionally to this view I may state that the results in the last paragraphs tend to remove one difficulty which was opposed to that hypothesis. This difficulty exists on one hand in the great difference between the absorbability of the cathode rays studied by Mr. Lenard and that of the x-rays and, secondly, in the fact that the transparency of these substances for cathode rays follows a law in relation to the density of the substance other than that for the transparency of x-rays.

With respect to the first point, two facts should be considered. (1) We have seen in paragraph 7 that there are x-rays varying greatly in absorbability, and we know from the investigations of Hertz and Lenard that the different cathode rays also differ from each other in their absorbabilities; thus if the "softest tube" mentioned in paragraph 7 produced x-rays whose absorption does not in any way approach that of the cathode rays investigated by Mr. Lenard, there exist without doubt x-rays of still greater and, on the other hand, cathode rays of still smaller absorbability. It therefore seems entirely possible that in further experiments rays may be found which, as far as their absorbability is concerned, form a link between one type of ray and the other. (2) We found in paragraph 4 that the thinner the layer of an irradiated substance is the smaller its specific transparency. Consequently, if we had used in our experiments plates as thin as those of Mr. Lenard, we might have found values for the absorption of the x-rays that would have been nearer those of Lenard's.

With regard to the different influences of the density of substances upon their absorption of x-rays and of cathode rays, it must also be stated that this difference is found to be the smaller the more easily absorbable the x-rays for this experiment are (paragraphs 7 and 8) and the thinner the irradiated plates (paragraph 5). Consequently the possibility must be admitted that this difference in the behavior of the two types of radiation, as well as the one mentioned previously, may be made to disappear by further experimentation.

Nearest in their absorbability are the cathode rays produced especially in very hard tubes and the x-rays, preferably emitted from the platinum plate, in very soft tubes.

10. In addition to exciting fluorescence x-rays also exert, as is well known, photographic, electrical, and other effects, and it is of interest to know to what degree these run parallel if the source of radiation is altered. I had to confine myself to a comparison of the two first mentioned effects.

The platinum-aluminum window is again very useful for this purpose. One of these was placed upon a wrapped photographic plate, a second was put in front of the fluorescent screen, and then both of them were placed at equal distances from the discharge apparatus. The x-rays had to traverse exactly the same media in order to reach the sensitive layer of the photographic plate and the barium platinocyanide. During the exposure I observed the screen and determined the window-number; after the photographic plate was developed, the window-number was also determined on it, and then both numbers were compared. As a result of such experiments no difference was observed when softer tubes were used (window-numbers 4 to 7); when using harder tubes it seemed to me as if the window-number on the photographic plate was slightly lower, but at most only one unit, than that determined with the fluorescent screen. However, this observation, although confirmed repeatedly, is still not entirely incontestable, since the determination of the high window-number on the fluorescent screen is rather uncertain. Absolutely certain, however, is the following result. If with the photometer described in paragraph 2 one adjusts a hard and a soft tube so as to produce equal brightness on the fluorescent screen and if one then substitutes a photographic plate for the screen, one observes, after the plate is developed, that that half which has been irradiated by the hard tube is considerably less darkened than the other half. The radiations which produce equal intensity of fluorescence have different photographic effects.

In judging this result one must not fail to consider that neither the fluorescent screen nor the photographic plate completely utilizes the impinging rays; both transmit many rays that can again produce fluorescence or photographic effects. The result given, therefore, applies only to the ordinary thickness of the sensitive photographic layer and of the layer of barium platinocyanide.

How very transparent the sensitive layer of the photographic plate is even for x-rays from tubes of medium hardness is proved by an experiment in which 96 films, one laid on top of the other, were placed 25 cm from the source of radiation and exposed for five minutes, the whole being protected against radiation from the air by a lead cover. A photographic effect can be clearly recognized even on the last one of them, while the first is hardly over-exposed. Induced by this and similar observations I asked several manufacturers of photographic plates whether it might not be possible to produce plates that would be more adapted to photography with x-rays than the ordinary ones. The samples obtained, however, were not serviceable.

I have had many opportunities, as already mentioned in paragraph 8, to notice that very hard tubes require a longer time of exposure than medium hard ones under otherwise identical conditions; this is understandable if one remembers the result mentioned in paragraph 9, according to which all examined substances were found to be more transparent to rays emitted by hard tubes than to those emitted by soft tubes. That even with very soft tubes a long exposure is again required may be explained by the smaller intensity of the rays emitted by them.

If the intensity of the rays is increased by increasing the primary current (see paragraph 9), the photographic effect is increased in the same degree as is the intensity of the fluorescence; in this as well as in the case discussed previously, in which the intensity of the radiation of the fluorescent screen was altered by changing the distance of the screen from the source of radiation, the brightness of the fluorescent screen might be—at least approximately—proportional to the intensity of the radiation. A general application of this rule, however, is not permissible.

11. In conclusion may I be permitted to mention the following details. In a properly constructed, not too soft, discharge tube the x-rays are emitted principally from a point 1 to 2 mm large at which the platinum plate is struck by the cathode rays; however, this is not the only point of emission: The whole plate and a part of the wall of the tube emit x-rays, although to a much smaller extent.

Cathode rays really travel from the cathode in all directions; but their intensity is significant only near the axis of the concave mirror, and therefore the most intense x-rays are produced at the point where the axis meets the platinum plate. If the tube is very hard and the platinum thin, a considerable quantity of x-rays is also emitted from the rear of the platinum plate and, as the pinhole camera shows, again mostly from a point lying on the mirror axis.

Also in these hardest tubes the maximum intensity of the cathode rays could be deflected from the platinum plate by a magnet. Some experiences with soft tubes led me to investigate once more and with better instruments the question of the magnetic deflection of x-rays; I hope to be able to report on these experiments soon.—

I have continued the experiments mentioned in my first communication on the transparency of equally thick plates cut from a crystal according to different directions. Plates of calcite, quartz, tourmaline, beryl, aragonite, apatite, and barite were examined. Again no influence of the direction upon the transparency could be detected.—

The fact observed by Mr. G. Brandes that x-rays can produce a light sensation in the retina of the eye, I have confirmed. In my record book there is also a note, written at the beginning of November 1895, according to which I noticed in a completely darkened room near a wooden door, on the other side of which a Hittorf tube was placed, a feeble sensation of light that spread over the whole field of vision when discharges were sent through the tube. Since I observed this phenomenon only once, I thought it was subjective; the reason that I did not see it again lies in the fact that later on, other less evacuated tubes without platinum anodes were used instead of the Hittorf tubes. The Hittorf tube because of its high evacuation produces rays of small absorbability and because of its platinum anode, which is struck by cathode rays, produces very intense rays, all of which favors the production of the sensation of light as mentioned above. I had to replace the Hittorf tubes with others because all of them were punctured after a very short time.

With the hard tubes now in use the Brandes experiment may be easily repeated. A description of the following experimental procedure is perhaps of some interest. If one holds a vertical metal slit, a few tenths of a millimeter wide, as close to the open or closed eye as possible, and if one then holds the head, enveloped in a black cloth, near the discharge apparatus, one observes after some practice a weak and not uniformly bright strip of light, which according to the position of the slit in front of the eye has a different shape: straight, curved, or circular. By a slow motion of the slit in a horizontal direction one can progressively make these forms pass into one another. An explanation of this phenomenon is easily found if one considers that the eyeball is intersected by a laminated beam of x-rays, and if one assumes that x-rays can produce fluorescence in the retina.—

Since the beginning of my work with x-rays I have made repeated efforts to obtain diffraction phenomena with these rays; several times when using narrow slits, and so forth, I also obtained phenomena whose appearance recalled diffraction patterns, but when the conditions of the experimental arrangements were altered in order to check the correctness of the explanation of these images as being produced by diffraction, it was refuted in every case, and I could frequently prove that the phenomena were produced in a manner entirely different from diffraction. I cannot recall one experiment on the basis of which I could safely be convinced of the existence of diffraction of the x-rays.

Würzburg, Physik. Institut der Universität.
March 10, 1897

# Chapter IX

# 1897-1923

O SCIENTISTS, Röntgen's three papers on *A New Kind of Rays*, have become a model for the presenting of results with brevity and simplicity; as in Luther's theses there is "never a word too little and never a word too much." As demonstrated in these famous communications, the salient features of his work were his perseverance and critical honesty in making observations and measurements. In approaching physical problems his acuity was inspired, and in solving them his thoroughness was relentless. Over and over again he devised new control experiments to prove beyond a doubt the absolute accuracy of the results obtained. He was always skeptical of any hypothesis not based upon sound experimental evidence. Yet, in spite of having an ascendant interest in experimental physics, he was convinced that good experimental results should be based upon indisputable theoretical reasoning. Before attacking a new problem he first crystallized the fundamental idea to make it amenable to experimentation. Building most of his experiments upon sound theory and logic, he seldom used higher mathematics, although he by no means underestimated the value of mathematics in certain problems. His discovery of x-rays, which secured him a place among great scientists of all time, was the consequence of his grasping the significance of an apparently unimportant effect of an unknown phenomenon and tracing this effect to its real cause.

Rarely in the history of science has a discovery been publicized as rapidly as was the discovery of x-rays or made such a deep impression upon the public. This may account for the many attempts to belittle Röntgen's accomplishment as an accidental finding or to ascribe it to another person, an assistant or a laboratory diener. Münsterberg, the great Harvard philosopher, silenced some of these voices of envy with:

Suppose chance helped. There were many galvanic effects in the world before Galvani saw by chance the contraction of a frog's leg on an iron gate. The world is always full of such chances, and only the Galvanis and Röntgens are few.

The attacks were often foolish and sometimes vicious and continued to appear periodically even after Röntgen's death. Two of the most violent were published in the *Münchner Post* in 1908 and in the *Zürcher Illustrierte* in 1935.

Röntgen's reticence, bordering upon bitterness with advancing years, was doubtless a defense against these attacks. In his first letter to Zehnder after the discovery he had already stated: "Let the envious chatter in peace." "It is almost as if I had to apologize for discovering the rays," he said later to a friend. In April 1921, two years before his death, he wrote to Mrs. Boveri:

... Zehnder has also heard the fable that I was not the first to notice the x-rays, but that an assistant or diener discovered them. What miserable envious soul must have invented this story? I also learned recently that it is said that I had nothing to do with the medical application of the rays and that the first photograph of a hand was made at the suggestion of Schönborn! However, Schönborn was the very one who was reluctant and even unfriendly in accepting the news of the discovery. I must admit that these lies affect me a great deal more than they should—I should remember: Where there is much light, there is also much shadow. ...

To Zehnder, he wrote at about the same time:

The infamous rumor that I did not discover the rays origi-
nated presumably in Quincke's institute in Heidelberg. Lenard
probably cultivated it. I was astonished while going over my
old letters to find some written by Lenard that show a very
friendly attitude toward me, which, however, stopped completely about the time Wien succeeded me in Würzburg and
I received the Nobel prize!

That Lenard's investigations on cathode rays formed a
significant basis for the discovery of x-rays, Röntgen always fully acknowledged. Yet Lenard's attitude toward
Röntgen, which was very friendly at the time of and for
some time after the discovery, eventually became hostile.
Lenard never used the term *roentgen rays* but preferred
*high-frequency rays,* and Röntgen's name was not mentioned in Lenard's Heidelberg institute.

During a controversy in 1929 over the discovery Lenard
intimated that perhaps more data in regard to the discovery
of high frequency rays might be revealed at a later date.
On August 18, 1929, he wrote to Glasser:

There is no doubt that the road to the discovery led over
my researches. At that time I was prevented by external circumstances from pursuing to my satisfaction in every direction
the great number of new phenomena that appeared in my
studies on cathode rays. But in my opinion, this is not yet the
proper time to express myself more thoroughly on the subject
than I did in my Nobel prize lecture. That would be only
biographical anyway, and what has already been said must
suffice for the judicious. With this I believe that I have done
everything that the history of science can expect of me on this
point at this particular time.

In all subsequent discussions Lenard maintained this
enigmatic attitude. The veil over the mystery was lifted

somewhat in 1935 when friends of Lenard made a concerted effort to prove that Röntgen used the Lenard tube in making the discovery. Although valuable historical material was then contributed, the evidence was not sufficient to alter the story of the discovery as given previously. Lenard's attitude filled Röntgen with great bitterness, and he looked in vain for Lenard's name among the German physicists who in 1905 dedicated a tablet in his honor, which was placed on the physical institute of Würzburg.

Having an intellectual honesty that characterized not only his work but also his attitudes, Röntgen had little patience with behavior actuated by selfish personal motives or hampered by personal prejudice; these interfered with the progress of science and were detrimental to the institution of such a person's affiliation. He could be gruff and even rude when aroused by nonproductive discussion with "people who are chiefly concerned with their own importance" or with those who were inclined to gild superficial knowledge with brilliant rhetoric.

Röntgen's greatness as a scientist was matched by his ability as an educator. He always stressed accuracy, intellectual honesty, open-mindedness, suspended judgment, search for true cause and effect, and critical evaluation, including self-criticism. Röntgen had little regard for popularization of science, and in referring to popular lectures, particularly on physics, he said:

Physics is a science which must be proved with honest effort. One can, perhaps, present a subject in such a manner that an audience of laymen may be convinced erroneously that it has understood the lecture. This, however, means furthering a superficial knowledge, which is worse and more dangerous than none at all.

He was not a brilliant lecturer, yet his lectures on experimental physics and his lecture demonstrations were presented with the thoroughness that characterized his research methods, and the same meticulousness was carried over into his laboratory exercises. Keen observation in all exercises was required of his students, whose ability he estimated by questions. He took great care in selecting subjects for doctors' theses. In emphasizing the need for self-reliance in the student he instructed his assistants, "Do not pamper the student, for it is useless. Let each find his own way out of difficulties." Many students, notably in medicine, who could not follow his classic presentation thought his lectures dry and stayed away. His examinations were dreaded and were not only difficult and never routine but always comprehensive. Even so, Röntgen had little respect for the examination and considered it a necessary evil. "The experience of life itself is the real test of capacity for any kind of profession," was his rule.

As director of the institute Röntgen considered it his duty to conserve and manage conscientiously the property and funds placed in his trust by the government. The same acceptance of responsibility was demanded of his subordinates, especially with regard to the collection of apparatus, care of it, and respect for its value.

The stature of the man made a tremendous impression upon the many scientists and laymen who visited him after his discovery. His distinguished features, penetrating gaze, and unassuming manner marked him as a man of extraordinary character and great dignity. In spite of his amiable and courteous nature a certain reticence or shyness, amounting almost to diffidence, was evident in his reception of strangers. With the passage of time, fame and this innate shyness erected a wall around Röntgen, which not only protected him from the selfish and curious but drove away many sincere persons. At times he even displayed a kind of mistrust, which often concealed a deep sympathy and understanding. Once convinced of the sincerity of another, he replaced the diffidence with warm friendliness. The few who entered into personal friendship with him kept faith with him until death.

Of the honors traditionally awarded distinguished men who become public figures, Röntgen received a lion's share. At the Kaiser's instigation a bronze statue, for which Röntgen posed with considerable reluctance, was sculptured by Felderhoff and erected on the Potsdam Bridge in Berlin. The most representative bust of Röntgen, however, was cast in bronze by the famous sculptor Hildebrand and placed in the Glyptothek in Munich in 1915. Numerous other busts were made after his death, none of which did him justice, with the possible exception of that by Georgii, which was unveiled at Munich University on the fifth anniversary of Röntgen's death, and of one other, sculptured by a Russian artist and erected in front of the roentgen institute of the University of Leningrad.

Recognition from abroad came with the golden Rumford medal from the Royal Society in London—this he sacrificed to the fatherland during World War I—, the Elliot-Cresson medal of the Franklin Institute in Philadelphia, and with his appointment as a Komtur of the Italian Crown.

Other universities honored him with efforts to obtain his services. In 1899 he refused to accept a call to the University of Leipzig, and in gratitude for his faithfulness he was honored by the title of Royal Geheimrat. However, on April 1, 1900, at the special request of the Bavarian government, he agreed to take over Lommel's new physical institute at the Ludwig-Maximilians-University in Munich.

Because of the wholly satisfactory living and working conditions in Würzburg, he was reluctant to make the change, and because of the adjustments and the administrative work consequent to the transfer to Munich, his time to do research was limited and the publication of some of his investigations delayed.

During the first few years in Munich Röntgen worked under various undesirable conditions, which the Ministry of Education ignored. The Röntgens also found it difficult to adjust to the stiff pattern of the official life after the carefree atmosphere of Würzburg. The Prince Regent accepted them graciously, but because of Röntgen's rejection of nobility some members of the court circle received them coldly. However, the cultural attractions of art, science, and music in the Isar-City appealed to them greatly, and despite many enticing offers Röntgen could not be persuaded to leave Munich.

Soon after moving to Munich Röntgen leased the *Gögerl* hunting ground at Weilheim, and in 1904 he bought a little cottage nearby to remodel into a hunting lodge. The furnishings were comfortable but unpretentious. An unusual feature was the large and excellent library. From the lodge the Röntgens had an incomparably beautiful view of the Bavarian Alps from the Algaeu to the Karwendel.

Although it was difficult for him to distinguish red deer from the green background of the forest because of green color-blindness, Röntgen was an excellent shot. This was even more remarkable because his vision in one eye had been considerably impaired by a childhood illness.

It was in small parties at the hunting lodge that Röntgen's true personality was revealed. After a day in the field the Röntgens and their guests would return to the lodge for a hearty dinner and would then linger around the table to swap hunting stories. In this conviviality Röntgen was at his best, and his deep laugh would ring out with complete freedom.

At Weilheim Röntgen kept a collection of pseudophysical instruments, which he demonstrated to his guests with great glee. One, called "simplified x-rays" and described by Margret Boveri, was a little oblong box containing four wooden blocks numbered 1 to 4. In Röntgen's absence from the room, these blocks were arranged in any sequence. After returning to the room Röntgen would hold an empty shell over the closed box and by looking into the shell would correctly name the sequence of the blocks. The trick was ingenious. The shell shown to guests was empty, and for it Röntgen substituted one in which was secreted a small compass. Tiny magnets were fastened to the wooden blocks, to the top of block 1, to the bottom of block 2, and to the left and right of blocks 3 and 4. The compass in the shell pointed north for the first block, south for the second, and so forth.

In Munich Röntgen resumed his researches on the physical properties of crystals and examined, partly in collaboration with the Russian physicist Joffé, their electrical conductivity and radiation influences upon them. Investigations on the nature and application of x-rays he always followed with keenest interest and trained many a leading radiation physicist in his laboratory. Some of the more important researches in his own institution included Angerer's on certain effect in the process of absorption of x-rays, Bassler's on polarization of x-rays, and Friedrich's on the distribution of intensity and quality of x-rays around the target.

After publishing the third communication on x-rays

Röntgen published nothing more for a number of years. In addition to his reluctance to prepare his data for publication, in later years his desire to investigate a subject most thoroughly often led him too far afield, necessitating re-vision and re-editing of the papers, which further delayed publication. Many important results he failed to publish because he feared the possible chance of an experimental error.

He continued to receive offers of high posts in universi-ties and awards for his distinguished contribution to science. In 1904 the Prussian government offered him the presi-dency of the Physikalisch-Technische Reichsanstalt in Ber-lin-Charlottenburg as successor to Kohlrausch, and in 1912 he was asked to accept van't Hoff's professorship at the Prussian Academy of Sciences in Berlin. Both offers he refused. On Christmas Day 1908 the Bavarian government gave him the title Excellency.

Two tributes came from the United States during these years. In 1900 he received the Barnard medal from Co-lumbia University, and in 1902 the Carnegie Institute in Washington offered him splendidly equipped laboratories to carry out special investigations of his choice. He refused the offer, but from that time he had an increasing desire to visit the United States. In 1912 he made plans, which never materialized, to accept the invitation of a number of Ameri-can scientific societies to attend their meetings and at the same time to visit his relatives in the United States.

In 1901, the year of its inception, the Nobel prize was given Röntgen for his contribution to physics. Contrary to his usual rule of not personally attending the awarding of an honor, he traveled to Stockholm in December 1901 and was welcomed on December 9 by the great Swedish chemist Arrhenius. Upon his arrival he wrote his wife, "The formal

Nobel Medal.

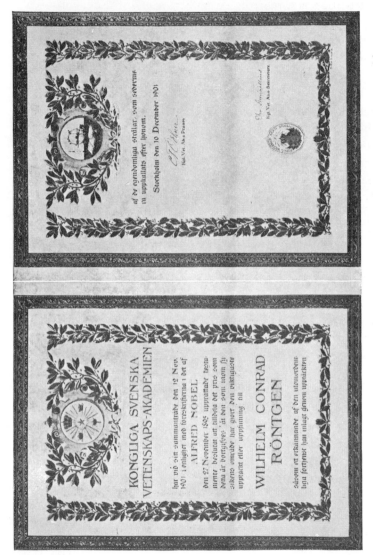

Nobel Diploma. "To Wilhelm Conrad Röntgen, in recognition of his eminent merits, proved by his discovery of the rays which today bear his name."

celebration is tomorrow evening at seven o'clock, followed by a supper, and for the next days there are many invitations from professors. I probably shall decline them and return soon. . . ."

On December 10, together with E. von Behring from Marburg and J. H. van't Hoff from Berlin, he received the diploma, gold medal, and prize from the hands of the Swedish Crown Prince in a ceremony at the Music Academy. He declined to give an official Nobel lecture, but at the impressive supper after the ceremony he spoke a few words of appreciation for the honor and said that this recognition would stimulate him to greater unselfish activity in science, which he hoped would be of benefit to humanity. The prize money he gave to the University of Würzburg to be used in the interest of science.

To celebrate the tenth anniversary of the discovery of the x-rays the roentgenologists Eberlein, Immelmann, Cowl, Gocht, Albers-Schönberg, Walter, Rieder, Grashey, and Köhler planned a Röntgen Congress in Berlin and invited Röntgen to be the honored guest. In writing to Zehnder, who lived in Berlin at that time, Röntgen said, ". . . Of course I shall not go to the Congress in Berlin which bears my name without my permission; I cannot understand how my friends could do such a thing to me. . . ."

The meeting opened on April 30, 1905 under the auspices of the Berlin Röntgen Society. On May 2 the Congress founded the German Röntgen Society and made Röntgen an honorary member. One of the congratulatory messages was from Röntgen:

Accept my sincerest thanks for the greetings sent to me by the executive committee in the name of the members of the Congress. Please let me assure you that I am filled with joy and admiration over the work which others, many of whom are now united in this congress, have derived from the discovery of the x-rays—Röntgen.

Much of the work of other physicists studying x-rays in the meantime was directed toward identifying their true nature. The answer was found in a department affiliated with Röntgen's own physics department at Munich University, the department of theoretical physics, established by Sommerfeld at Röntgen's insistence in 1906. There in 1912 Friedrich and Knipping, working on a suggestion of von Laue, obtained the first x-ray diffraction patterns.

Friedrich personally showed Röntgen the apparatus and the confirmatory pictures. After listening to Friedrich's detailed explanation of the method and after examining the experimental evidence thoroughly and critically, Röntgen finally declared that he could find no experimental error, but he was hesitant about accepting the pictures as incontestable proof. He congratulated Friedrich upon his fine experimental result but added, "You know, I do not think that these are interference phenomena; they look different to me." That these young men might have been successful where he had failed in attemping to discover the true character of x-rays recalled his first communication in which he had written, "I have often looked for interference phenomena of x-rays. . . ." Soon thereafter, however, he became convinced that these pictures actually demonstrated interference phenomena of x-rays, which classified them with visible light, ultraviolet light, and all other sections of the electromagnetic spectrum as transversal waves.

In these experiments von Laue, Friedrich, and Knipping had made the first fundamentally new contribution to the knowledge of the nature of x-rays since Röntgen had published his papers seventeen years before, that is, with the

possible exception of the discovery of secondary cathode
rays by Dorn or of polarization of roentgen rays by Barkla.
Confirmatory evidence was contributed in Röntgen's own
institute by the studies of Wagner and Glocker on the mono-
chromatic character of the diffracted rays and by the famous
experiments of Bragg, father and son, in England.

The outbreak of World War I and the ultimate defeat
of Germany affected Röntgen deeply. He was particularly
distressed that many spheres of activity, notably pure
science, should not have been but were affected by prejudices engendered by the conflict. Yet in embracing the cause
of his country he displayed relatively little animosity.

In addition to the burden of increased responsibilities in
teaching and research incident to the war emergency, Röntgen was depressed by his wife's chronic illness. Shortly after
moving to Munich in 1900 Mrs. Röntgen began to com-
plain of attacks of severe pain, and although she felt perfectly well between attacks, her condition grew worse
yearly. A diagnosis was made of renal colic, but because of
her advanced age the surgeons refused to operate. More
and more the Röntgens' activities were restricted because
of her condition. She could endure only the shortest jour-
ney and was permitted little pleasure except that from
working with her flowers. During the war years Röntgen
gave her as many as five morphine injections a day to
relieve the intense pain, and he could never leave her for
more than a few hours. With this tremendous drain on
his energy, he was often melancholy.

On October 31, 1919 Mrs. Röntgen died in an attack
of the illness. The memory of their wonderful companion-
ship for almost fifty years intensified Röntgen's loss, and in
his loneliness he would read important news items to her

picture and pretend that she still shared his thoughts. On the anniversaries of her birth and death he paid special tribute to her memory.

Röntgen's own health had generally been good, although in 1913 he had suffered lung hemorrhages and frequent vertigo after strenuous exercise. During the same year he had had a rather serious ear operation from which he had quickly recovered.

In the critical period of the war and the postwar years the lodge at Weilheim became more of a retreat than ever. To supplement the meager rationed food portions Röntgen planted a garden of vegetables and one year even raised a pig. He was extremely conscientious in seeing that his household adhered to rationing regulations and often measured the daily rations of fat, meat, flour, and sugar with a scale. The meats and vegetables that were not used immediately were canned. But in one luxury he did not restrict himself. That was smoking. He was very fond of strong Dutch tobacco and preferred to smoke it in cheap clay pipes, which he cleansed by placing them in the fire. "These clay pipes," he said, "really taste best, which proves that not everything has to be expensive to be good."

He followed the events of the war with profound interest and was able to foresee with an astounding insight many of the bitter consequences that eventuated for Germany. Having many contacts with relatives and friends in neutral countries, he was able to judge the situation more accurately than if he had been entirely dependent upon German news releases. Although he was not blinded to weaknesses inherent in certain German institutions and practices, the sudden collapse of the German navy and army in 1918 and its sad sequelae came to him as a surprise. "The loss of Alsace, especially of Strassburg, is particularly tragic to me, for I was present at the time the university was reopened under tremendous universal enthusiasm, and it was there that I spent one of the most beautiful and productive periods of my life," Röntgen wrote to Margret Boveri on November 19, 1918.

Although he was apprehensive about the outcome of Germany's chaotic political affairs immediately after the war, he maintained an open-minded attitude toward the efforts of the new social democratic government. With a certain prophetic outlook he stated:

. . . One may well expect from this government the best that can be hoped for. Of course, it is a question whether it can hold its own in the future, at least during the most difficult period of transition, or whether ultraradical influences may gain the upper hand. . . .

And early in 1923, a few weeks before his death, Röntgen again looked into the future:

Martial law was declared in Bavaria yesterday, and no public meetings are permitted without government permission. It is said that this regulation is directed against the National Socialists, who under Hitler—perhaps a second Mussolini (?) —caused some disturbance. I wonder whether it is not too late to lay the ghosts that have already been called forth?

One serious consequence of the lost war was the collapse of the German currency. Röntgen, who had a very comfortable fortune invested to a large extent in Dutch, Swiss, American, and Italian securities, lost most of his savings by changing his foreign investments into German War Loan Bonds. In a letter to Mrs. Boveri, dated April 10, 1920 and written from Munich, he said:

. . . There (at the bank) I was disagreeably surprised to learn that I must turn in most of my American and Italian securities to the government and that the final date for doing this is the fifteenth of this month. Not complying with this law would mean a severe fine. . . . I had some difficulty in convincing the official that I had had no intention of defrauding him, that is, of keeping possession of this money a secret. . . . Now I am going to receive a great deal of German money, and I do not really know what to do with it. . . .

Röntgen adapted himself to these new conditions and even began to buy some stocks, although he was amused that at his age he should become interested in the stock market. Most of his fortune, however, was destroyed by the inflation shortly after he died.

During these years great strides were being taken in the direction of the systematic application of radiation in the treatment of disease. In collaboration with Friedrich, Krönig, the great Freiburg gynecologist at the newly founded radiologic institute of the Frauenklinik at Freiburg University, published a monograph in 1918 on the physical and biological foundations of radiation therapy. Röntgen followed this work closely, as well as Friedrich's subsequent report on the great value of x-rays in the treatment of many diseases, notably cancer.

Although x-ray treatments had been given for years, initially by Freund in Vienna, Williams and Grubbe in the United States, and by many others, this was the first systematic determination of correct radiation quantities and qualities by new and accurate physical methods for controlled therapeutic results. This contribution combined with extension of the method to treatments with gamma rays of radium by Friedrich's pupil, Glasser, had, according to Krönig, opened a new era in radiation therapy. To Friedrich's laboratory after the close of the war came visitors from all over the world to learn methods of measuring radiation dosages. Among these were Gray and Hopwood from London, Puga from Lisbon, Nauta from Batavia, Maisterra from Madrid, and Fujinami from Tokyo. From the United States there were Hickey of Detroit, Case of Chicago, Crane of Kalamazoo, Schmitz of Chicago, and Coolidge of Schenectady, and from Canada Richards of Toronto.

The standardization of diagnostic and therapeutic x-ray exposures was greatly advanced by the development of a hot filament cathode tube, which was remarkably stable and permitted regulation of voltage and current at the tube. The tube was the product of Coolidge's researches in the laboratories of the General Electric Company in Schenectady. The stability of the new tubes was due to Coolidge's process for producing ductile tungsten, used in the manufacture of targets for the tubes. X-ray generators had also been continuously improved and had become instruments of precision.

One of the students who had received training in Röntgen's laboratory expressed Röntgen's reaction to these developments of his discovery with these words:

The constantly growing significance of the discovery filled Röntgen with a certain pride in having given humanity so great a blessing, and occasionally one could see a quiet gleam of happiness in his eyes when one overcame one's shyness and reported to him some new development in connection with the rays.

It was greatly consoling to Röntgen to know that his discovery had contributed so much to the amelioration of pain and the saving of lives during the war, and that this knowledge in counteracting to a small degree the ravages of war was benefiting mankind as a whole. In acknowledgement of his contribution to the war effort, he received the order of the "Iron Cross on the white-black ribbon in full recognition of the great value of your rays in restoring wounded soldiers." This, with a congratulatory letter from Field Marshal von Hindenburg, arrived on Röntgen's seventieth birthday, which he was spending with his gravely ill friend Boveri in Oberstdorf.

Years later a letter reached him which quoted an American doctor, R. C. Beeler, from Indianapolis:

We were in the trenches near Toul, when we heard that the roentgenologists in the German hospitals were celebrating Professor Röntgen's birthday. The American radiologists appreciate the discovery of the professor as much as the Germans do. We drank with French cognac to the health of the old German professor. They shall not say that we are narrow-minded or prejudiced. We recognize the celebrity of this man and only wished that old Professor Röntgen could have heard us.

Röntgen had survived not only his wife but also many of his friends, and in his last years he was often very lonely. He became professor emeritus in the spring of 1920. Two small laboratories in the physical institute were retained for his use, and he also continued to serve as Conservator of the Physical Metronomical Institute of the Munich Academy of Science.

One pleasure left him was a lively correspondence with his friends. "It is especially beneficial for the lonely old man who has lost his life companion and who must live in these sad times to know that there are still people who remember him in the most intimate way," he wrote in 1921. In his letters he commented on the rapidly changing social and economic scene but more often recalled amusing and vivid vacation days of the past. Frequently he recounted the details of spring vacations in Italy and Corfu, the upper Italian lakes, the Riviera, or Santa Margherita-Ligure. Occasionally he and Mrs. Röntgen had traveled as far as Cairo on these southern jaunts, but they had usually directed their course to Caddenabia at Lake Como. Many of his reminiscences were of his birthdays spent there with Krönlein, Stöhr, Hofmeier, Hitzig, and Boveri at the Hotel Bellevue. In past summers they had gone to Pontresina in the Engadin after two weeks at some Swiss resort at a lower altitude, preferably Lenzerheide, Flims, or Rigi-Scheidegg. Pontresina always recalled the *Weisses Kreuz* and his old friends von Hippel, Lüders, Ritzmann, and von Gaffky. Often Röntgen relived the yearly return from Pontresina to Munich, frequently by horse carriage in order to enjoy the beauties of the Swiss mountains at leisure. They had always hired the same driver, Schmidt, brother of the Bishop of Chur, Schmidt von Grüneck. And in the winter his memory turned to New Years celebrated in Davos in Switzerland, where he had tobogganed down the icy path from the Schatzalp.

In his last years Röntgen's nostalgia for the Swiss mountains was intensified. During the war, because of responsibilities at the physical institute and Mrs. Röntgen's ill health, he had had no opportunity to return to Switzerland, and after the war the inflation of the German mark had made it impossible for him to finance the trip.

In 1921, after letters and finally a telegram from Wölfflin in Basel in which he urged Röntgen to spend the summer in Switzerland as his guest, Röntgen returned to the Engadin. In his letter of appreciation to Wölfflin he wrote:

This is what I wanted to see once more before I die. The roaring mountain stream is for me the symbol of potential power. . . . One of the most beautiful episodes was the delightful trip from Tiefenkastel to Lenzerheide; there past and present were united in one brilliant event. . . . I still prefer to leave the well worn paths and to hike over stick and stone. . . . If I ever should be missed do not look for me on the main road. . . .

The following year, again at Wölfflin's invitation, he paid his last visit to the Engadin, and on August 9, 1922 he made his last hike into the mountains, up the beautiful Fexthal, starting from Sils Baselgia between the Silvaplaner and the Silser lakes. He rested on a rock at the Marmorei and looking back saw for the last time the two blue lakes below him, the deep cut valley to the right leading to St. Moritz, the chain of snow mountains directly before him, Piz Materdell, Piz Lagrev, and Piz Julier.

Leaving the Engadin on August 13 Röntgen and Wölfflin took the Rhätische railroad to Tiefenkastel and from there traveled by "postauto" to Lenzerheide. There they stayed for another two weeks at Cantiene's inn, and occasionally Röntgen talked about religion with the Bishop of Chur, Schmidt von Grüneck. Toward the end of August Röntgen returned to Munich without being able to stop in Zürich for a last visit with friends and a final glimpse of his Alma Mater.

The holiday had renewed his energy, and with a new enthusiasm he returned to work. However, the walk to the laboratory became a strain, and he was easily fatigued. His vision was not keen enough for him to make accurate observations, but he kept at his work until a few days before his death.

On December 6, 1922 he wrote to his cousin Louise:

Dear Louise—I have not received an answer to my letter of April 15 of this year but now Christmas-time and the New Year approaches, and that is the time when one likes to get in touch again with those loved ones who are still alive. As usual

Röntgen monument in Lennep.

I can be rather sure to receive some news from you within the near future, provided of course that you are feeling well and are in the mood for writing. I hope that this is the case, but especially at our age, this is very often uncertain. I frequently ask myself how you are, and then I should like to have an answer.

I am feeling rather well. My hearing and sight have decreased considerably and other signs of age have appeared, but I am still rather active and have a good appetite. Memory and ability to work are considerably decreased, and loneliness lies heavily upon me. . . .

At my wedding you no doubt met the children of Coo Boddens (my cousin). I still have some correspondence with one of them (Betry). Outside of that I have no connections any more with Holland.

Since I am a government employee, I receive a pension, which increases somewhat with the rise of prices of living and am thus relatively secure if I live very moderately, but I must continuously practice more and more economy. Just think of it: One pound of bread costs 67 marks; one pound of meat 300 to 400 marks; one pound of butter about 1400 marks, and a simple suit of clothes about 150,000 to 200,000 marks. For one dollar one can get over 8000 marks.

You can imagine that under these circumstances I ordinarily cannot travel in countries that have a high exchange rate, but still I was able this summer to spend three weeks in Switzerland. That was a wonderful time! During this time, I was with people who live under normal conditions. Close to the place where we stayed for some time, there is a famous bathing resort, St. Moritz; in the guest list of one of its greatest hotels I found a "Mr. Ernst Roentgen with Governess and Maid, U.S.A." Do you know who this apparently very wealthy namesake could have been?

And now, my dear Louise, I wish you a very Merry Christmas; begin the New Year in good health and with courage. Sincerest regards, Your cousin, W. C. Röntgen.

In November 1922, a few days after the anniversary of
his wife's death, he traveled with Mrs. Boveri to Giessen
to make arrangements with the city authorities to have the
Röntgen family graves permanently taken care of. Such an
arrangement was impossible to make at that time because
of the inflated German mark.

After his wife's death and the death of Boveri, Röntgen
and Mrs. Boveri and her daughter Margret looked to each
other for support and release from loneliness. The Christmas
holidays they spent together in Würzburg. The postwar
Christmas of 1922 was celebrated with gifts of food from
Mrs. Boveri's friends in the United States and delicacies
from Röntgen's friends in Switzerland. They made the
occasion festive with gay Christmas decorations and old
Christmas music.

Many evenings Röntgen spent pouring over documents
and papers. One tribute that he often re-read was the
address sent him by the Prussian Academy of Sciences in
Berlin on the occasion of the fiftieth anniversary of his
doctorate.

Esteemed colleague! The fiftieth anniversary of the happy
day on which you started your scientific career is also a day of
rejoicing for our Academy. We cannot let it pass without
giving expression to the joyful pride that among our number
we can count you, whose name is greatly praised by all mankind.

In your youth a kind Fate led you into Kundt's laboratory,
and permitted you to complete your education under the sur-
veillance of this master of experimental art.

Your first investigation on the ratio of specific heats of gases
furnished an immediate example of your keen, critical, and
precise workmanship. Having accomplished this work, but
still under the influence of your teacher, you soon entered other
fields of investigation which allowed the individuality of your

scientific personality to emerge clearly. Your great gift of finding new methods was shown in the simple and yet ingenious method of measuring the conductivity of heat in crystals by studying the form of the *Hauch Figur*.

Your astounding talent for construction manifested itself thoroughly in the investigations, made in collaboration with August Kundt, on the electromagnetic rotation of the plane of polarization in gases. In this classical work, you succeeded in observing and measuring quantitatively in several gases the effect which had been sought in vain by Faraday.

Again you showed the same skill in overcoming experimental difficulties in the many investigations made with your pupils on the influence of pressure upon compressibility, capillarity, viscosity, and the refraction of light of various bodies. As a testimony of this valuable work, one need only consider your theory on the constitution of liquid water which has proved to be so extremely fruitful.

By employing a new original method, you brought the old discussion between John Tyndall and Gustav Magnus on the absorption of heat rays by water vapor to a final decisive solution. You tackled a question of fundamental significance in your investigations of the electrodynamic effect of a dielectric moved through an homogeneous electrical field. The fact that you succeeded in observing with certainty the extremely small effect predicted by the Maxwell theory again is an indication of your highly developed art of experimentation.

All of these investigations, among which must also be mentioned your comprehensive systematic examinations of the pyro- and piezoelectricity of crystals, are worthy of securing you an honored place among the leading physicists of Germany. However, these prominent scientific accomplishments, as compared with your great discovery of 1895, pale as the stars before the sun. Probably never has a new truth from the quiet laboratory of a scientist made triumphal progress so quickly and universally as has your epoch-making discovery of these wonderful rays. The expectations as to the theoretical and practical value of the new discovery were tremendous, but these have been far surpassed by the actual developments.

The history of science shows that in each discovery usually there is a peculiar connection between merit and chance, and many a man not entirely familiar with the facts may be inclined to ascribe a preponderant part to luck in this particular case. But whoever has penetrated into the singularity of your scientific personality understands that this great discovery was destined to belong only to you, to the investigator free from all prejudices, who combined perfect experimental art with the greatest scrupulousness and conscientiousness.

The three communications in which you described the wonderful properties of the new rays belong to the classical works of physical science for their modest form, for their essential brevity, and for their masterful presentation. The value of discernment contained in your discovery has opened a new era in our science, which constantly arrives at more gratifying results and reaches out to higher goals.

The eminent practical significance of the new rays, which you recognized at once but which you, in your noble unselfishness, have left to others to develop practically, was revealed in a most striking manner during the World War. One can say with complete authority that the fruits of your scientific investigations have spared life and limb to hundreds of thousands of poor wounded soldiers, both friend and foe. Thus you are not only esteemed by physical science as its immortal master but also by all humanity as its benefactor.

May the joyful satisfaction of having contributed so greatly to the furtherance of our knowledge and to the benefaction of suffering mankind help you on this day of jubilee to overcome the distress which we all feel over the collapse of our beloved fatherland. May you live to see the dawn of better times. This is our sincere wish.

(Signed) THE PRUSSIAN ACADEMY OF SCIENCES

On the tenth anniversary of his discovery his physics

colleagues had bestowed upon him another honor to which in his old age he often turned:

ESTEEMED COLLEAGUE: Ten years have passed since you gave to humanity the great discovery of your rays. With it you have broken a new path for our science, on which it has advanced to great achievement within a very short time. As a consequence of your discovery, almost every year has brought to light new and fundamental phenomena.

In the name and at the request of the German physicists, we wish to express our gratitude by placing on the Physical Institute of the University of Würzburg, the site of your great discovery, a tablet with the following inscription:

"In this building, W. C. Röntgen discovered the rays which are called by his name."

These diplomas were carefully locked away with many others in a bookcase in his studio. It was his wish that all his diplomas, addresses, medals, and so forth, referring to his scientific work, should be given to the University of Würzburg to be preserved in the physical institute. "Perhaps they are of historical interest," he said. "They probably must be gone over and a proper selection be made." A number of his original tubes, including the pear-shaped Hittorf-Crookes tube, and his original induction coil he had already given to the Deutsches Museum in Munich. According to his will his scientific and personal writings and correspondence were to be burned by the executors. This wish was followed to the letter, and as a consequence material that would have been invaluable in rounding out the picture of his life was destroyed.

At about this time the Swiss Röntgen Society placed a tablet on the house in Zürich at No. 7 Seilergraben, where he had lived as a student, with this inscription:

Wilhelm Conrad Röntgen, the discoverer of the rays named

after him, lived here 1866 to 1869 when a student at Federal Technical High School.

Gastrointestinal distress, which Röntgen had suffered at intervals for the past year, returned. By unmistakable signs he diagnosed his illness as carcinoma of the intestines. Von Müller, director of the medical clinic of the university, did not confirm the diagnosis and told Röntgen that there was no cause for alarm. However, to relieve himself during attacks of labored breathing, Röntgen ordered an oxygen flask with the proper attachments from the physical institute.

Röntgen died on February 10, 1923. This announcement the press gave to the world:

> Heute früh halb 9 Uhr verschied nach kurzer Krankheit im 78. Lebensjahr Se. Excellenz Geheimrat Professor
>
> **Dr. Wilhelm Conrad Röntgen.**
>
> In tiefster Trauer
> die Verwandten und Freunde.
>
> München, den 10. Februar, 1923.
> Die Einäscherung findet am Dienstag, den 13. Februar 1923 vormittags 10 Uhr im östlichen Friedhofe statt.

"The whole German nation mourns at the bier of its great son," wrote the Minister of Interior. The funeral cortège assembled prominent scientists from all parts of Germany and neighboring countries, who came to pay their respects to a great confrère and benefactor of mankind. Tributes were spoken by his peers, Planck, Sauerbruch, von Göbel, von Drygalski, Ruland, Wien, and von Müller. Schubert's *Litanei* and Haydn's *Largo* ended the service. Röntgen's ashes were placed besides those of his wife and parents in the Giessen cemetery on November 10, 1923.

The full life that was Röntgen's is to be rejoiced. His greatest gain was in that he was able to give—to humanity, to science, to his students, and to his friends, and from each he took in proportion—gratitude, fame, respect, and good fellowship.

Röntgen museum in Lennep.

# Scientific Papers of W. C. Röntgen

1. On the Determination of the Ratio of the Specific Heats for Air. Ann. Physik u. Chem., 141: 552 (1870).
2. Determination of the Ratio of the Specific Heats for Constant Pressure to Those for Constant Volume for Several Gases. Ann. Physik u. Chem., 148: 580 (1873).
3. On Soldering Platinum-Plated Glasses. Ann. Physik u. Chem., 150: 331 (1873).
4. On Conducting Discharges of Electricity. Ann. Physik u. Chem., 151: 226 (1874).
5. On a Variation of the Sénarmont Method for the Determination of the Isothermal Areas in Crystals. Ann. Physik u. Chem., 151: 603 (1874).
6. On an Application of the Ice Calorimeter for the Determination of the Intensity of Sun Radiation. (With Exner.) Wien. Ber. (2), 69: 228 (1874).
7. On the Ratio of Cross Contraction to Longitudinal Dilation of Caoutchouc. Ann. Physik u. Chem., 159: 601 (1876).
8. A Telephonic Alarm. Nature (Lond.), 17: 164 (1877).
9. Communication on a Few Experiments on Capillarity. Ann. Physik u. Chem., N.F. 3: 321 (1878).
10. On an Aneroid Barometer with Mirror and Scale. Ann. Physik u. Chem., N.F. 4: 305 (1878).
11. On a Method for the Production of Isothermals on Crystals. Z. Kryst., 3: 17 (1878).
12. On Discharges of Electricity in Insulators. Göttinger Nachr., 390 (1878).
13. Demonstration of the Electromagnetic Rotation of the Plane of Polarization of Light in Vapor of Carbon Disulphide. (With Kundt.) Münch. Ber., 8: 546 (1878). Ann. Physik u. Chem., N.F. 6: 332 (1879).

14. Supplement to the Paper on the Rotation of the Plane of Polarization in Carbon Disulphide. (With Kundt.) Münch. Ber., 9: 30 (1879).

15. On the Electromagnetic Rotation of the Plane of Polarization in Gases. (With Kundt.) Ann. Physik u. Chem., N.F. 8: 278 (1879). Münch. Ber., 8: 148 (1879).

16. On the New Relation between Light and Electricity Found by Mr. Kerr. Ann. Physik u. Chem., N.F. 10: 77 (1880). Ber. d. Oberhess. Ges. f. Nat. u. Heilk., 19.

17. On the Electromagnetic Rotation of the Plane of Polarization in Gases. 2. Communication. (With Kundt.) Ann. Physik u. Chem., N.F. 10: 257 (1880).

18. On the Changes in Form and Volume of Dielectrics Caused by Electricity. Ann. Physik u. Chem., N.F. 11: 771 (1880).

19. On Sounds Produced by Intermittent Irradiation of a Gas. Ann. Physik u. Chem., N.F. 12: 155 (1881). Ber. d. Oberhess. Ges. f. Nat. u. Heilk., 20.

20. A New Method for the Measurement of Absorption of Rays by Gases. Ber. d. Oberhess. Ges. f. Nat. u. Heilk., 20: 52 (1881).

21. The Changes in Double Refraction of Quartz Caused by Electric Forces. Ann. Physik u. Chem., N.F. 18: 213, 534 (1883). Ber. d. Oberhess. Ges. f. Nat. u. Heilk., 22: 49.

22. Observation on the Communication of Mr. A. Kundt: On the Optical Properties of Quartz in the Electrical Field. Ann. Physik u. Chem., N.F. 19: 319 (1883).

23. On the Thermo-, Actino-, and Piezo-electrical Properties of Quartz. Ann. Physik u. Chem., N.F. 19: 513 (1883). Ber. d. Oberhess. Ges. f. Nat. u. Heilk., 22.

24. On an Apparatus for the Lecture Demonstration of Poiseuille's Law. Ann. Physik u. Chem., N.F. 20: 268 (1883).

25. On the Influence of Pressure upon the Viscosity of Liquids,

Notably Water. Ann. Physik u. Chem., N.F. 22: 510 (1884).

26. New Experiments on the Absorption of Heat through Water Vapor. Ann. Physik u. Chem., N.F. 23: 1, 259 (1884) Ber. d. Oberhess. Ges. Nat. u. Heilk., 23.

27. Experiments on the Electromagnetic Effect of Dielectric Polarization. Math. u. Naturw. Mitt. a. d. Sitzgsber. Preuss. Akad. Wiss., Phys.-Math. Kl., 89 (1885).

28. On Compressibility and Surface Tension of Liquids. (With Schneider.) Ann. Physik u. Chem., N.F. 29: 165 (1886).

29. On the Compressibility of Diluted Salt Solution and of Sodium Chloride. (With Schneider.) Ann. Physik u. Chem., N.F. 31: 1000 (1887).

30. On the Electrodynamic Force Produced by Moving a Dielectric in a Homogenous Electric Field. Math. u. Naturwiss. Mitt. a. d. Sitzgsber. Preuss. Akad. Wiss., Phys.-Math. Kl., 7 (1888).

31. On the Compressibility of Water. (With Schneider.) Ann. Physik u. Chem., N.F. 33: 664 (1888).

32. On the Compressibility of Sylvin, Rock Salt, and Potassium Chloride Solutions. (With Schneider.) Ann. Physik u. Chem., N.F. 34: 531 (1888).

33. On the Influence of Pressure upon Refraction Coefficients of Carbon Disulphide and Water. (With Zehnder.) Ber. d. Oberhess. Ges. f. Nat. u. Heilk., 28: 58 (1888).

34. Electrical Properties of Quartz. Ann. Physik u. Chem., N.F. 39: 16 (1889).

35. Description of the Apparatus with Which the Experiments on the Electrodynamic Effect of Moving Dielectrics Were Made. Ann. Physik u. Chem., N.F. 40: 93 (1890).

36. Some Lecture Demonstrations. Ann. Physik u. Chem., N.F. 40: 109 (1890).

37. On the Thickness of Coherent Oil Layers on the Surface of Water. Ann. Physik u. Chem., N.F. 41: 321 (1890).

38. On the Compressibility of Carbon Disulphide, Benzol, Ethylic Ether, and Several Alcohols. Ann. Physik u. Chem., N.F. 44: 1 (1891).

39. On the Influence of Pressure on the Refractive Index of Water, Carbon, Disulphide, Benzol, Ethylic Ether, and Several Alcohols. (With Zehnder.) Ann. Physik u. Chem., N.F. 44: 24 (1891).

40. On the Constitution of Liquid Water. Ann. Physik u. Chem., N.F. 45: 91 (1892).

41. Short Communication on Experiments Dealing with the Influence of Pressure on Some Physical Phenomena. Ann. Physik u. Chem., N.F. 45: 98 (1892).

42. On the Influence of Heat of Compression on the Determination of the Compressibility of Liquids. Ann. Physik u. Chem., N.F. 45: 560 (1892).

43. Method of Producing Pure Surfaces of Water and Mercury. Ann. Physik u. Chem., 46: 152 (1892).

44. On the Influence of Pressure on the Galvanic Conductivity of Electrolytes. Nachr. Ges. Wiss. Göttingen, Math.-Physik. Kl., 505 (1893).

45. The History of Physics at the University of Würzburg. Würzburg, 1894, p. 23.

46. Note on a Method to Measure Differences in Pressure by Means of Mirror and Scale. Ann. Physik u. Chem., N.F. 51: 414 (1894).

47. Communication on a Few Experiments with a Right Angle Glass Prism. Ann. Physik u. Chem., 52: 589 (1894).

48. On the Influence of Pressure upon the Dielectric Constant of Water and Ethyl Alcohol. Ann. Physik u. Chem., N.F. 52: 593 (1894).

49. On a New Kind of Rays. Sitzgsber. Physik.-Med. Ges.

Würzburg, 1895, cxxxvii. Ann. Physik u. Chem., N.F. 64: 1 (1898).

50. On a New Kind of Rays. Second Communication. Sitzgsber. Physik-Med. Ges. Würzburg, 1896, xi. Ann. Physik u. Chem., N.F. 64: 12 (1898).

51. Further Observations on the Properties of the X-rays. Math. u. Naturw. Mitt. a. d. Sitzgsber. Preuss. Akad. Wiss., Physik-Math. Kl., 392 (1897). Ann. Physik u. Chem., N.F. 64: 18 (1898).

52. Communication. Physik Z. 5: 168 (1904).

53. On the Conductivity of Electricity in Calcium Spar and the Influence of X-rays upon It. Sitzgsber. Bayer. Akad. Wiss., Math.-Physik. Kl., 37: 113 (1907).

54. Friedrich Kohlrausch. Sitzgsber. Bayer. Akad. Wiss., Math.-Physik. Kl., 40. Schluss-H., 26 (1910).

55. Determinations of the Thermal Linear Coefficient of Expansion of Cuprite and Diamond. Sitzgsber. Bayer. Akad. Wiss., Math.-Physik. Kl., 318 (1912).

56. On the Conductivity of Electricity in Several Crystals and on the Influence of Irradiation upon Them. (Partly in cooperation with A. Joffé.) Ann. Physik, 4, F. 41: 449 (1913).

57. Pyro- and Piezo-electrical Investigations. Ann. Physik, 4, F. 45: 737 (1914).

58. On the Conductivity of Electricity in Several Crystals and on the Influence of Irradiation upon Them. (Partly in cooperation with A. Joffé.) Ann. Physik, 4, F. 64: 1 (1921).

# Chronology
# W. C. Röntgen's Life

1845—March 27.   Born at 287 Poststrasse in Lennep, Rhine Province, Germany.

1848—May 23.   Röntgen family moved to Apeldoorn, Holland.

1862—Dec. 27.   Entered Utrecht Technical School, Holland.

1865—Jan. 18.   Entered University of Utrecht to audit courses as a student of mechanical engineering.

Nov. 16.   Moved to Zürich, Switzerland as a student of mechanical engineering at the Federal Polytechnical School.

1866.   Met Anna Bertha Ludwig, who was born on April 22, 1839, in Schwamendingen, Switzerland and who became Röntgen's wife in 1872.

1868—Aug. 6.   Graduated as mechanical engineer from the Polytechnical School of Zürich, Switzerland.

1869—June 22.   Received Ph.D. degree from the University of Zürich and became assistant to August Kundt, professor of physics.

1870.   In the capacity of assistant, Röntgen followed Professor Kundt to the Julius-Maximilians-University at Würzburg.

1872—Jan. 19.   Married Anna Bertha Ludwig of Zürich in Apeldoorn, Holland.

April 1.   Followed Professor Kundt to the University of Strassburg as his assistant.

1873—Oct. 3.   Röntgen's parents moved from Apeldoorn to Strassburg to live near their children.

1874—March 13.   Became *privat-dozent* in physics at the University of Strassburg.

1875—April 1.   Became professor of physics and mathematics

at the Agricultural Academy at Hohenheim in Württemberg.

1876—Oct. 1. Returned to the University of Strassburg as
associate professor of theoretical physics.

1879—April 1. Became professor of physics at the Hessian
Ludwigs-University in Giessen.

1884—June 12. Röntgen's father died at Giessen.

1886. Declined offer of chair of physics at the Friedrich-
Schiller-University in Jena.

1887. The Röntgens took Josephina Bertha, Mrs. Röntgen's
six year old niece, into their home and adopted her when
she was twenty-one.

1888. Declined offer of chair of physics at the University of
Utrecht.

Aug. 8. Röntgen's mother died at Bad Nauheim.

Oct. 1. Became professor of physics at the Julius-
Maximilians-University in Würzburg.

1894—May. Professor August Kundt died at his summer
home near Lübeck.

June. Röntgen acquired an improved Lenard cathode ray tube from Müller-Unkel in Braunschweig for
experimental studies.

Became rector of the University of Würzburg.

1895—February. Declined offer of chair of physics at the
Albert-Ludwigs-University in Freiburg i. Br.

Nov. 8. Röntgen discovered the first effects due to
*a new kind of rays.*

Dec. 28. Submitted the manuscript of a paper *On
a New Kind of Rays, a Preliminary Communication* to
the secretary of the Würzburg Physical-Medical Society for publication in its *Sitzungsberichte.*

1896—Jan. 1. Sent reprints of his first communication and
copies of his x-ray pictures to his colleagues Exner,
Kohlrausch, Lummer, Poincaré, Schuster, Voller, Warburg, and Zehnder.

Jan. 4. Röntgen's first x-ray pictures were exhibited

at the physical institute of Berlin University on the occasion of the fiftieth anniversary celebration of the Berlin Physical Society.

Jan. 5. First newspaper story of the discovery of x-rays appeared in the Vienna *Presse*.

Jan. 6. News of the discovery was cabled all over the world.

Jan. 13. Röntgen demonstrated the x-rays before Kaiser Wilhelm II and Kaiserin Victoria at the Berlin castle, he was decorated with the Prussian Order of the Crown, II Class.

Jan. 23. Lectured on his discovery of x-rays before the Physical-Medical Society at the physical institute of the University of Würzburg.

Jan. 30. Declined to lecture before the German Reichstag and several scientific societies.

March 3. Received honorary degree of Doctor of Medicine of the University of Würzburg.

March 9. Submitted the manuscript of his second communication *On a New Kind of Rays, Continued* to the secretary of the Würzburg Physical-Medical Society.

March 10. Left for Sorrento to escape the flood of honors and invitations.

April 16. Made honorary citizen, Lennep, Röntgen's birthplace.

April 20. Received Royal Order of Merit of the Bavarian Crown. This order carried with it personal nobility. Röntgen accepted the decoration but refused nobility.

May. Corresponding member, Prussian Academy of Sciences, Berlin.

November. Corresponding member, Bavarian Academy of Sciences, Munich.

Nov. 30. Rumford Gold medal, Royal Society, London.

Baumgaertner prize, Vienna Academy.

Honorary member, Naturforscher-Gesellschaft, Freiburg i. Br.

Honorary member, Société Scientifique Antonio, Alz. Mexico.

Honorary member, Physical Society, Frankfort o.M.

Honorary member, Chester Society of Natural Sciences.

Corresponding member, Société Nationale des Sciences et Mathématiques, Cherbourg.

Corresponding member, Wissenschaftliche Gesellschaft, Göttingen.

1897—March 10. Submitted his third communication *Further Observations on the Properties of X-rays* to the Prussian Academy of Sciences, Berlin for publication in its *Sitzungsberichte*.

Elliot-Cresson medal, Franklin Institute, Philadelphia.

Prize Lacaze, Académie des Sciences, Paris.

Mattencei medal, Rome.

Honorary member, Swiss Naturforscher Gesellschaft.

Honorary member, Physical-Medical Society, Erlangen.

Honorary member, Röntgen Society, London.

Honorary member, Société des Médecines Russes, Petersburg.

Honorary member, Société Impériale de Médecine, Constantinople.

Honorary member, Society of Former Students of Federal Polytechnical School, Zürich.

Corresponding member, Reale Accademia di Geographici, Florenz.

Member, American Philosophical Society, Philadelphia.

1898. Prize, Otto-Wahlbruch-Stiftung, Hamburg.

Honorary member, New York Medical Society.

Corresponding member and bronze plaque, Reale Accademia dei Lincei, Rome.

Corresponding member, Reale Instituto Veneto di Scienze.

Nonresident member of the Société Hollandaise des Sciences, Harlem.

1899.   Received offer of chair of physics at the University of Leipzig but declined. Received the title of *Royal Geheimrat* from the Bavarian Government.

Diploma, University of Zürich.

Corresponding member, Cataafsch Genootschap, Rotterdam.

Nonresident member, Royal Academy of Sciences, Stockholm.

1900—April 1.   Became professor of physics and director of the physical institute at the Ludwig-Maximilians-University in Munich.

Grosskomturkreuz of the Royal Order of Merit of the Bavarian Crown.

Order of Merit of Saint Michael, I Class.

Silver medal of Prince Regent Luitpold.

Komtur of the Order of the Italian Crown.

Member of the Maximilian Order for Sciences with Decoration.

Barnard medal, Columbia University, New York.

Honorary member, 1. German Academy for Physical Dietetics Therapy, Hamburg.

Honorary member, Society of Physicians, Munich.

Nonresident member, Academy of Medicine, Paris.

Member, Academy of Sciences, Munich.

1901.   Nobel prize in physics, Stockholm.

Honorary member, Physical Society, Stockholm.

1902.   Received invitation of Carnegie Institute in Washington to use its laboratory for special work but did not accept.

Honorary member, Instituto de Coimbra.

1903.   Honorary member, Philosophical Society, Cambridge.

Honorary member, Berlin Röntgen Society.

Corresponding member, Reale Accademia dei Scienze, Turin.

1904. Declined offer of presidency of Physikalisch-Technische Reichsanstalt in Berlin-Charlottenburg.

Honorary member, Society of Physicians, Vienna.

Röntgenstrasse, Cologne.

1905—March 27. Address of German physicists Boltzmann, Braun, Drude, Ebert, Graetz, Kohlrausch, Lorentz, Planck, Riecke, Warburg, Wien, Wiener, and Zehnder on the occasion of the tenth anniversary of the discovery of x-rays, which was also Röntgen's birthday; plaque at physical institute of University of Würzburg.

Honorary member, Society for the Encouragement of Arts, etc., London.

Honorary member, Medico-Chirurgical Society, Edinburgh.

Honorary member, German Röntgen Society.

1906. Folder, board of trustees, Deutsches Museum, Munich.

Honorary member, Royal Institute of Great Britain.

1907. Nonresident member, Societa Italiana delle Scienze, Rome.

Member, Royal Academy of Sciences, Amsterdam.

1908. Honorary member, Society of Physicians, Stockholm.

Honorary member, German Medical Society, New York.

Title of Excellency bestowed upon Röntgen by the Prince Regent of Bavaria.

1909. Röntgenstrasse, Würzburg.

Honorary citizen, Weilheim.

1910. Honorary member, German Society of Neurologists.

Honorary member, Berlin Medical Society.

1911. Röntgenstrasse, Halle.

Honorary member, Society of Physicians, Smolensk.

Order Pour le Mérite for Science and Art.

1912. Declined offer of professorship at Prussian Academy of Sciences, Berlin.

Russian diploma, Odessa.

Planned to visit relatives in America.

Address, life membership, Deutsches Museum, Munich.

1913.   Honorary member, German Surgical Society.

Honorary member, Swiss Röntgen Society.

1914.   Honorary member, New York Röntgen Society.

1915—March 27.   Röntgen spent his seventieth birthday quietly with his best friend, Boveri, who was seriously ill, in Oberstdorf, Bavaria; on the next day he had an audience with the King of Bavaria in Munich.

Addresses and honors on the occasion of Röntgen's seventieth birthday:

Würzburg Medical Faculty

Röntgen Foundation

University of Giessen

Philosophical Faculty, University of Würzburg

Physikalisch-Technische Reichsanstalt

University of Strassburg

Annalen der Physik

Röntgenstrasse, Munich.

Iron Cross, II Class.

1918.   Honorary degree of Doctor of Engineering, Technical High School, Munich.

1919—June 22.   Fiftieth anniversary of Röntgen's doctorate in philosophy. He received the following honors:

Fiftieth doctorate diploma, University of Zürich.

Address of Prussian Academy of Sciences, Berlin.

Helmholtz medal in bronze.

Helmholtz medal in gold.

Honorary member, German Physical Society.

1920.   Röntgen retired from the University of Munich and became professor emeritus.

March 27.   Röntgen's seventy-fifth birthday. He received the following honors:

Honorary degree of Doctor of Natural Sciences, Johann Wolfgang Goethe-University, Frankfort o.M.

Honorary member, Bonn Röntgen Society.

Honorary member, Frankfort Röntgen Society.

Honorary member, Society for Natur and Heilkunde, Dresden.

Commemorative tablet placed on the house of Röntgen's birth and an address by the mayor of the city of Lennep.

Nonresident member, Prussian Academy of Sciences, Berlin.

Nonresident honorary member, Vienna Academy of Sciences.

Member, Strassburg Scientific Society, Heidelberg.

Röntgenstrasse, Lennep.

Röntgenstrasse, Weilheim.

Honorary citizen, Würzburg.

1921. Honorary member, Society and Friends of the Friedrich Wilhelms-University of Bonn.

Honorary member, Nordisk Foerening f. med. Radiologi.

Honorary academician, University of Bonn.

Corresponding member, Phys. Oekonom. Society.

1922. Commemorative tablet placed on the house Röntgen lived in during his Zürich college years, Swiss Röntgen Society.

1923—Feb. 10. Röntgen died at Munich.

Nov. 10. Röntgen's ashes were put to rest in the family grave at Giessen.

1928—Dec. 9. Röntgen Memorial Room opened at the physical institute, University of Würzburg.

1928—July 27. Röntgen bust, sculptured by Georgii, unveiled at the University of Munich.

1930—Nov. 30. Röntgen monument unveiled in Lennep.

Röntgen-Realgymnasium, Lennep.

1932—June 18. Röntgen museum opened in Lennep.

# Bibliography

The content of this biography of Röntgen is chiefly based upon

Glasser, Otto: Wilhelm Conrad Röntgen and the Early History of the Roentgen Rays; with a chapter, Personal Reminiscence of W. C. Röntgen, by Margret Boveri.

> Berlin, Springer, 1931
> London, Bale Sons & Danielsson, 1933
> Springfield, Ill., Thomas, 1934

and upon numerous letters, documents, and other communications obtained subsequent to that publication. All contributions on Röntgen's life appearing before 1933 are acknowledged in the above biography.

Valuable material has been obtained from the following publications, which have appeared since 1933.

Donaghey, J. P.: Reminiscences of Röntgen, Radiog. & Clin. Photog. 10:2, 1934.

Evers, G. A.: Wilhelm Conrad Röntgen in den Niederlanden. Acta radiol. 16:88, 1935.

Freund, L.: Wie es zur Entdeckung der Röntgenstrahlen kam. Wien. med. Wchnschr. 83:403, 1933.

Gerlach, W.: W. C. Röntgen, der Forscher und sein Werk in der Auswirkung fuer die Entwicklung der exakten Naturwissenschaften. Strahlentherapie 47:3, 1933.

Glasser, Otto: First Observations on the Physiological Effects of Roentgen Rays on the Human Skin. Am. J. Roentgenol. 28:75, 1932.

————: What Kind of Tube Did Röntgen Use When He Discovered the X-rays? Radiology 27:138, 1936.

————: The lift of Wilhelm Conrad Röntgen as Revealed in His Letters. Scient. Monthly 45:193, 1937.

Günther, P.: Röntgen als Briefschreiber. Angewandte Chemie 50:77, 1937.

Henkels, P.: Wilhelm Conrad Röntgen zum Gedenken. Deutsche Tieraerztl. Wchnschr. 41:81, 1933.

Rössler, O.: Zur Entdeckung der nach Röntgen benannten Strahlen. München. med. Wchnschr. 16:631, 1935.

Sarton, G.: The Discovery of X-rays. Isis 26:349, 1937.

Schinz, H. R.: Röntgen und Zürich. Acta radiol. 15:562, 1934..

Schmidt, F.: Ueber die von einer Lenard-Fensterroehre mit Platinansatz ausgehenden Roentgenstrahlen. Physikal. Ztschr. 36:283, 1935.

Stark, J.: Zur Geschichte der Entdeckung der Roentgenstrahlen. Physikal. Ztschr. 36:280, 1935.

Thurstan-Holland, C.: X-rays in 1896. Liverpool Med. Chir. J. 45:61, 1937.

Weil, E.: Some Bibliographical Notes on the First Publication of the Roentgen Rays. Isis 29:362, 1938.

Wien, M.: Zur Geschichte der Entdeckung der Roentgenstrahlen. Physikal. Ztschr. 36:536, 1935.

Zehnder, L.: Persoenliche Erinnerungen an W. C. Röntgen und ueber die Entwicklung der Roentgenroehren. Helvet. Physica acta 6:608, 1934.

———: Persönliche Erinnerungen an Röntgen. Acta radiol. 15:557, 1934.

———: W. C. Röntgen's Briefe an L. Zehnder. Zürich, Rascher, 1935.

Zimmern, A.: Röntgen et la découverte des rayons x. Presse méd. 22:1, 1932.

The illustrations were obtained through the courtesy of the physical institute of the University of Würzburg and the Röntgen Museum, Lennep.

Since the first publication of DR. W. C. RÖNTGEN in 1945 the following papers and books have appeared:

Glasser, Otto: Strange Repercussions of Röntgen's Discovery of the X-Rays. Radiology 45:425, 1945.

———: Fifty Years of Roentgen Rays. Radiog. & Clin. Photog. 21:58, 1945.

———: Röntgen and von Laue. In 'Les Inventeurs Célèbres'. Paris, Mazenod, 1950.

———: Wilhelm Conrad Röntgen als Physiker. Röntgen-Blätter, 5:147, 1952.

————: The Human Side of Science. In 'The Doctor Writes.' Grune & Stratton, New York, 1954.

————: Erinnerungen eines alten Freundes des Deutschen Röntgen-Museums. Röntgen-Blätter, 7:426, 1954.

————: The Sixtieth Birthday of the Roentgen Rays. Bull. Acad. Med. of Cleveland, 9:5, 1955.

A number of additional articles and books on Röntgen have been published by various authors; they are mostly similar to previously published material. New viewpoints are presented in the following articles:

Boveri, M.: Wilhelm Conrad Röntgen. In 'Die Grossen Deutschen.' Berlin, Propyläen-Verlag, 1956.

Etter, L. E.: Post-War Visit to Röntgen's Laboratory. Am. J. Roentgenol. 54:547, 1945.

————: Some Historical Data Relating to the Discovery of the Roentgen Rays. Am. J. Roentgenol. 56:220, 1946.

Streller, E.: Beitrag zur Geschichte verschiedener Röntgen-büsten. Röntgen-Blätter, 8:147, 1955.

The following books on Röntgen, published in recent years, are based almost exclusively upon the previously published biographies on Röntgen by Zehnder and Glasser, and the references given therein. Some of these recent books are written in a romantic novel-like style.

Carstensen, R.: Die tödlichen Strahlen. Kiel, Neumann & Wolff Verlag, 1954.

Dessauer, F.: Wilhelm C. Röntgen. Die Offenbarung einer Nacht. Frankfurt a.M. Josef Knecht Verlag, 1951.

Hartmann, H.: Wilhelm Conrad Röntgen. In 'Schöpfer des Neuen Weltbildes. Grosse Physiker unserer Zeit.' Bonn, Athenäum Verlag, 1952.

Neher, F. L.: Röntgen, Roman eines Forschers. München, Braun & Schneider, 1936.

————: Blick ins Unsichtbare. Reutlingen, Bardtenschlager Verlag, 1956.

Sieper, B.: Der Ruf der Strahlen. Ein Biographisches Röntgenbildnis. W.-Elberfeld, Hans Putty Verlag, 1946.

Unger, H.: Wilhelm Conrad Röntgen. Hamburg, Hoffmann & Campe Verlag. 1949,

# Author Index

# Subject Index